EARTHWORM FEEDS AND FEEDING
by
Charlie Morgan

EARTHWORM FEEDS AND FEEDING

by

CHARLIE MORGAN

Thirteenth Edition - 1979

© Copyright 1961 by Charlie Morgan.

© Copyright Renewed 1989 by Charlie Morgan

Revised Edition
© Copyright 1970 by Charlie Morgan

All rights reserved.

Printed in the United States of America

No part of this book may be reproduced or reprinted in any form without written permission of the copyright owner. Copyright secured under International and Pan American copyright conventions.

published by -

SHIELDS PUBLICATIONS
Eagle River, WI 54521

ISBN 0-914116-02-9

PREFACE

In the pages that follow I have tried to point up a few of the worst pitfalls that lie in wait for the unwary—and sometimes the experienced and properly suspicious—earthworm grower.

My success will be measured to a great extent by the curiosity aroused, rather than the information propounded. My biggest problem has been to lead the average earthworm farmer to think for himself. The beginner wants hard and fast rules to follow, and assurance that following those rules will guarantee his success. The successful worm farmer who is doing well merely wants to know what other people would do in some of the situations that have confronted him.

There are no hard and fast rules to follow in the worm business; at least none that I have been able to classify and enumerate. Each situation must be met with the information, experience, and materials at hand.

No book of any size could define the actions and materials to use in all the different situations because the writer of the book could not possibly have met and solved all the conditions in their various combinations: chemical, seasonal, climatic, and thermal. Raising worms is more or less a matter of experience: Either the experience of the grower or of an author whose instructions he is following. These experiences continue on and on forever, so there is no possibility of anyone ever setting them all down in print for others to be guided by.

Follow the rules and guides that have been classified. Give your problem some thought. Do not depend on some distant grower to solve your problems for you. Learn to be independent. Worm growers are individualists—be one yourself. If a situation gives you more trouble than you think it should, outline it for me and I will do what I can with information and encouragement.

One thing: Please do not expect me to outline the contents of the books for your guidance—read them...And when you write, please send a stamp or a stamped envelope. You would be surprised at the number of letters that come to my office with no name, no return address, and especially no zip code number. Usually we do not bother to answer the latter at all.

Good luck, good sales, and good fishin.

EARTHWORM FEEDS AND FEEDING

INTRODUCTION

Methods of feeding earthworms and types of feed to use in the feeding of different types of worms have been a sort of mystery to most people who buy worms for bait or breeders, and are even to some extent obscure to many people who, by dint of trial and error, have been more or less sucessful worm growers.

Everyone wants to find a better product and a more efficient way to use it. A few people attempt to learn by active research and experiment with various feeds and methods and worms. But more popular, and a lot easier and cheaper, is the method of reading and asking questions. Any method has its drawbacks and frustrations. Much time, plus many dollars, is involved in experimental research. Most of us have little time and even less money that we can afford to expend in a work which may go on for many years, as it has in my own case.

Asking and looking also requires a great deal of time and money because of the necessary travel involved, and is usually not very rewarding in information because of reticence on the part of those asked, or the inability of the looker to recognize anything of particular significance when it is seen.

Thus the best, in fact the only way, open to the average worm grower or would-be worm grower is to lay the ground work of knowledge by reading enough of the subject to permit of experimentation directly applied to the problem at hand. By reading we avoid the time consuming preliminaries of applying unknown quantities of unknown materials to a more or less unknown, or at best but vaguely known, problem.

No book can give detailed instructions for meeting and overcoming all the perils that may arise in all the combinations of circumstances in which they may occur in a geographical area as large as that covered by American commercial worm growers. The best book can only give case histories of those problems encountered and solved, or those which are likely to be encountered in given situations, or are likely to become hazards to a majority of our readers.

Next to selling the product of your worm beds, feeding is probably the most important because good feeding habits produce better worms and prevent most of the problems that cause large losses of worms in the beds. Selling is most important for without an outlet there is no profit in the crop, and without profit we cannot continue, but selling is a primary effort that is encountered by anyone who produces anything, no matter what, and is solved according to the conditions governing demand and distribution.

Some worm growers seem to have no trouble selling the product of their labors, while others, whose product may be as good or better, will labor over it and fail. It has been disproven in the worm business that the man who has the best product will be most successful—up to now the men who have the best promotion methods have been most successful—although those with a better product and an honest approach are probably happier with the whole thing and are probably able to sleep better at night.

Experience is the best teacher is an old saying. True: but it is cheaper and easier to learn from the experience of others whenever it is possible. Most of us cannot afford the time to learn by our own experience, and find it better to buy the experience of others in the form of books.

There is no need to say that scientific terms will not be used here. No scientific subject can be adequately written about without the use of the terms and expressions developed by scientists, of whom I modestly claim to be one. Such language has its limitations, for the scientist as well as the layman. But some things simply cannot be said without the use of the word or words necessary to express that particular thought. Therefore I will be limited only by my knowledge of and ability to use the proper word or phrase to express my thought.

As an example: An earthworm's "head" or the "head" of an earthworm. Certainly the earthworm has no head as the term is commonly applied, so what should it be called? The front end? The leading end? It does not always lead! The only acceptable word then should be the scientific one: The anterior of the worm. Surely there is nothing difficult or confusing about a worm's anterior—or the other end, the posterior.

Much valuable information concerning types of earthworms and their habitat and food is covered in various scientific works, and the ability to translate the terms and reduce the subject matter to common popular language and thought is a valuable asset to anyone interested in the subject, even though a person trained in science may spend more time in searching a dictionary for meanings than in absorbing the knowledge imparted by the text of the book or passage.

Practically all writings on the subject of earthworms, scientific or otherwise, agree that the common species of worms will feed on and can subsist on practically any vegetable matter when the growth is stopped and the cells become separated by bacterial action or oxidation. The chemical action of sun, water, or man-made additives will break down vegetable cells, as will the by-products of bacterial or fungal growth. Some bacteria produce acid which remains in the habitat while others produce gases that are a result of the acids acting on plant cells.

Naturally each type of earthworm has its favorite food and habitat. But they are extremely adaptable to less favorite environment when required by circumstances. Nature, previous environment, and evolution of the species has fitted certain worms to live in certain types of material, and at least one kind (Chaetogaster) that is a true earthworm but will nevertheless live only in decaying flesh or the residue of decaying flesh in the earth. Other kinds require large quantities of organic matter (Eisenia), while still others live in sand or clay soil with a minimum of food in the form of decaying organic material.

One of the bitter ironies often encountered by worm growers is that some types, notably the more common "red wiggler" and other domesticated types, excluding the African nightcrawler, will grow larger in material in which they must be forced to live. A good example of this is sewer sludge. Worms of all kinds absolutely refuse to live anywhere near sewer sludge if there is any other form of food available, yet they will grow very fat in a short time if they are forced to live in a mixture of sludge and other materials--that is all but the African nightcrawler, we have noted that the African does no better in sewer sludge--indeed not as well, as in milder beddings and feeds. A footnote to this is that the worms must be taken out when they reach optimum condition because they begin to lose weight very rapidly and soon die; which only goes to show that an inexperienced worm is often smarter than an inexperienced human researcher.

Reference will be made often in this book about the use of antibiotics, or "wonder-drugs" as part of the adventure of searching for something better. Some writers point out that neither we or the worms had ever heard, until a few years ago, of any antibiotic such as Aureomycin, Gallimycin, Penicillin, and others. Quite true. And until a few years ago, certainly within the memory of many, none of us had ever heard of television, jet planes, air conditioning, or electronic hearing aids. Does the fact that a thing is new mean that we should refuse to use it or take advantage of its being? I do not believe so.

Does the fact that man only recently discovered the germ killing ability of a soil derivative also mean that nature may not have used the same thing since the beginning of life on our planet? Certainly not!

All antibiotic and antigerm cultures of the mycin type are developed from cultures taken from the soil. In fact the discovery of these germ killing fungi, after penicillin, came from scientists who wondered what happened to all the billions of bacteria dumped into and on the soil by men and beasts. Their inquiries led to the now quite obvious fact that nature, in her inimitable way, had long ago anticipated and solved the problem by developing various molds, or mushrooms of a microscopic size, that enveloped the germs and killed them, or gave off an acid that killed by dissolving the chetin, or shell, that covers their body.

Maybe we had never heard of antibiotics but the worms had, and made full use of the presence of germ killers in the soil. In fact, one of the functions of earthworms is to mix and till the soil to encourage the growth of such fungi, while discouraging by aeration the continued growth of bacteria.

All manures contain natural antibodies developed and used by nature in the intestines of animals, including man, for the specific purpose of controlling bacterial development. It is not accident that practically all manure, wherever found, is infested with fat, happy earthworms. Worms know what they like.

It is significant that one of the best early antibiotic cultures came from Egypt, where the soil probably supports more men and animals--and earthworms--per acre than any more heavily cultivated area on earth. Other good ones later came from India, where the soil is contaminated by men and animals to a greater degree than may be found anywhere else on earth.

When worms are taken from their natural environment, where bacterial growth is controlled by natural antibodies, and placed in material that is not only highly conducive to worm development and propogation but is also highly conducive to bacterial development, it is only common sense to add small amounts of natural antibiotics in the form of cultured commercial mycelium, even if indirectly, to living things. Bacteria themselves are harmless, usually, even in the bloodstream. It is the by-products and wastes in the form of acids and gases that cause harm.

No one in his right mind would use antibiotics to fatten worms, or implies that they may be useful as food. Had worms ever known of chicken laying mash or cotton seed meal or crushed oats before they met them under man-controlled conditions? Germs, bacteria, are about the only thing that can cause an increasing amount of acid in the beds. Germs manufacture small amounts of acid to dissolve their own food. The acid (and gases resulting from the action of this acid on cellulose, sugar, starches, etc., contained in the bedding and feed) manufactured by bacteria may or not, depending on the amount, be harmful to other organisms in the bedding--but too much of it will kill your worms.

Nature furnished us the antibiotics in the form of mycins (even if it did take 4.000 years of civilization for us to figure it out) to combat these germ and fermentation cells, thus reducing the amount of acid and the rate of breakdown of the bedding in our worm beds. One would be foolish not to use the bounties that nature provides, even if it is in the processed concentrated form of a drug.

At the risk of belaboring this IMPORTANT point:: Every one has heard of penicillin. It cures many diseases that were a few years ago incurable. Penicillin is derived from a mold, or fungus, is one of the "wonder drugs" and is an antibiotic product of some of the very mold that you may see growing in your worm beds, your home, or your garden. It is a very close relative of the yeast that is used by bakers to raise the bread you eat. Nature supplies this control for the benefit of every living creature.

Limestone, crushed or pulverised, is also used to control the acid that is formed in the presence of bacteria. Obviously, if enough of the acid is destroyed, the bacteria will also be destroyed because they are enabled by the acid to digest and convert to energy those amino-fractions that as a whole serve as the units of structure of the proteins which can be converted by replacing one of the atoms in the carboxyl group with one of the nitrogenous group. These converted to reconstructed proteins are essential to all animal metabolism, and the bacteria cannot live without them. Neither can the earthworm live without converted protein.

An accumulation in the bedding of too many converted proteins often causes "sour crop" a condition in which worms swell to the bursting point, usually at the clitellum, or protein poisoning where the worms squirm about on top of the bedding, get stringy, and swollen places appear as knots on the body of the worm. Or the worms may simply "hibernate" at the bottom of the bed and refuse to come up and eat food placed for them, which will aggravate the condition as the uneaten food decays. Or the worms may merely turn white and die in the bedding. Such symptoms are usually heralded by the appearance of mites in either their white or red form (in the white form they appear almost as a mold, or fungus; in the red form they are very active--the white form is one of the larval stages through which they progress from the egg to the adult red mite), white worms (Enchaetraeous, sometimes called nematodes), and other pests which inhabit only that bedding which has been overfed or otherwise carelessly or improperly cared for, and in which as a result, bacterial acid has formed.

Crushed limestone, which is composed largely of calcium carbonate, or oyster shell, which has a high calcium content, or purified calcium carbonate, may be more easily available and less bothersome for the average user than antibiotics. Calcium carbonate in its pure form is not a natural element but it is formed in high percentage with some other natural or deposited forms, such as chalk, marble, limestone, talc, etc.

Gallimycin poultry formula is an excellent antibiotic for use in affected worm beds and is commonly available through any poultry supply house or salesman. If it is not, write directly to Abbott Laboratories, Veterinary Division, North Chicago, Ill. and ask for the name of your nearest dealer, or a direct price quotation for mail delivery to you.

Any modern or "new fangled" product that will serve the

advancement of your earthworm project, whether that project is new or established, certainly should be used to the fullest extent of its capacity to be useful to you.

****

PRODUCT LIMITATIONS

First in the problem of feeding earthworms is reduction of those things most readily available and cheapest not only to buy, but to transport and process. This varies so widely in this extensive country and in its many climates as to be impossible to even list without the use of much more space than is available here. Let us start with the premise that any vegetable matter that is not inherently poison, of a volative nature, or pitchy, can be used for worm feed and or bedding. Leaves of all spicy plants such as bay, eucalyptus, magnolia, etc. should be avoided as should the needles of pines, firs, and cedars. This is usually true because they rot slowly, but eucalyptus or bay leaves contain an aromatic oil that may kill the worms or drive them from the bed.

The earthworm has a fairly complicated digestive apparatus but it does not include teeth nor does it include acid as a strong glandular secretion, as in mammals, that can be used to convert proteins and carbohydrates. The worm can and will ingest particles larger than is generally thought, but these are swallowed to a crop, resembling that of a chicken, where it is acted upon by enzymes and bacteria controlled by a very weak calcium solution. The composition of the substance called enzyme is still much of a mystery but it is known that it acts on organic and cellular matter to produce lypase, a digestive substance also found in cows and other ruminants. Protease, a form of incomplete protein found in vegetables and grain, and among other things, rennin, a form of vegetable "milk" which in reality is a "chyme" of converted starches and proteins. All these chemical actions and reactions are necessary to transform inorganic forms to organic (carbon) forms, of which all earth forms are composed. Each organism is a group of millions or billions of tiny cells, each of which must continually be added to or replaced in order that metabolism may continue. When food is lacking, metabolism and cell growth are reduced.

Remember that few or none of the activities of bacterial action that cause the conversion of the substances named above originate in the stomach of the earthworm. The worm farmer must start all this by his actions in bringing the proper materials together in the proper proportions under the proper conditions, which he must control in order to achieve the most affective results. Earthworms can and will starve to death, die of malnutrition, in the presence of feed if bacteria, ordinary germs, do not process it for them.

On the other hand the process must not be allowed to progress beyond certain fixed limitations. When the worm is forced to ingest food that contains too much acid and where bacterial action is very strong, the small calciferous glands contained in the worm's digestive system may not be sufficient to cope with the problem, with the result that fermentation continues in the worm's crop and gizzard and "protein poisoning" results. The crop swells and in many cases actually bursts. Or peritonitis may result when severe wounds in the intestinal tract become infected with disease bacteria.

Many earthworm growers, in fact most growers, learn enough by reading to be able to bring the materials together and many of them even bring them together in the proper proportions, but there, in many cases, they stop. They either neglect to add materials as needed, such as feed and water, or they fail to control either by proper observation and test.

From the crop of the earthworm the food goes to the gizzard, also very much like that of a chicken in that it is lined with a layer of "skin" composed largely of pepsin, or whose walls secrete pepsin in a form to further digest and combine proteins and starches.

The gizzard also contains and retains to some extent, grit, small particles of sand, and other hard substances that are rolled and ground together in such a manner as to grind softer substances into a mass of microscopic particles. These are the "teeth" of the earthworm--mere stones enclosed in a spherical muscle.

From the gizzard the food is transported by muscular action (the earthworm has no peristaltic effect in the intestines, as do mammals) into and through the digestive tract where such food particles as can be absorbed by the body are passed through the walls together with liquids. The proteins, sugars, and other cellular particles are taken up into the blood which flows along the worm's largest artery adjacent to its intestine, and is thus transported to the muscles and tissue where it is needed. The waste matter brought from the tissues is passed through cylia, a crude form of kidney, two of which are located in each segment, to the outside of the skin in the form of slime, which acts as a lubricant for the worm as it makes its way through rough soil and, incidentally, makes the worm hard to hold after it is captured.

All of which means that the earthworm, having digestive organs of the most rudimentary sort, and hardly any digestive fluids, must depend on external bacteria—decay germs—for "predigestion" of its food. This is accomplished by the acid excreted by bacteria to digest their own food. This acid takes several forms, most common of which is alcohol, plus various gases generated when the acid oxidizes cellular matter. The acid and gases must be maintained in more or less balance in order to not be harmful to the worms and not to become the breeding place of the various pests described in another chapter.

The farmer must of course use some judgement in the use of counteracting agents otherwise all bacteria may be destroyed. In warm weather this can be started again merely by feeding and watering. Alkalinity is not a usual situation to encounter but it can happen. It first affects the worms as would dehydration, or drying of the worm. The worm loses color and weight. They shrink and may later turn very dark and become lifeless. This is brought on only by extreme carelessness or by the use of an improper type of lime product. It can hardly happen with antibiotics as they are dissolved and washed out by continued normal watering.

The only reasonable method of maintaining accurate control is the use of soil acidity testers. These are cheap and can be had from mail order houses, or your county agent can instruct you.

The bedding should be about 1½ points on the acid side. There is room for variation here. Other things happen to worm beds and earthworms than just acid, and in many cases the symptoms may be the same to external observation. Sometimes bedding that kills the worm or drives them out over the side or to the bottom of the bed may not test acid. This is usually caused by too much or too little water, contaminated (salty or heavily chemically treated) water, and in many cases in recent years by weed killers and insecticides that have been used on crops which were subsequently used in worm beds. Only observation will establish the trouble here, or perhaps complete chemical analysis at your state or other university laboratory. Sewer sludge in some cases is seriously contaminated with detergent cleansers and may kill worms in a very short time, or so inhibit them that they will not feed.

One of the most troublesome problems in all my case histories was one caused by a weed killer. The weed killer had been used to deaden or kill the foliage to prepare cotton for a mechanical picker. The unwashed burrs and waste was subsequently picked up and used by a worm grower who had been in business for years. taking the cotton burrs (motes, etc.) from the same cotton gin and using them for bedding in his worm pits. In this one year he lost almost all of his worms---and lost them just when the best of the season was coming on, the spring. It was simple to solve the problem once we learned (by state university chemical analysis) what was causing it. He still uses the cotton motes but they are now all allowed to stand in the weather at least six months or are carefully spread out and thoroughly washed down with a heavy stream of water.

Never use ordinary garden or agricultural lime, quicklime, slacked lime, plaster lime, or any of the salt or soda products in the beds. All these things have their place in the "experimental wormery" but most will form dangerous and highly toxic gases that may kill your worm or destroy the egg capsules. Either crushed limestone, crushed oyster shell, powdered chalk, dolomitic limestone, or one of the antibiotics will be available for use. I may say here that dolomitic limestone is not always effective or safe. Much of it has a natural phosphorous radical that may damage or kill your worms. Check the analysis, usually printed on the bag or on attached tag and if it shows more than about .04% phosphorous it is not safe to use in large quantities though it may be used in small quantities, not more than one half pint to ten square feet of bedding.

Heavy spring rains often cause damage in worm beds other than ordinary flooding, and sometimes worse. Gases being formed in the bedding are sometimes trapped there by the heavy water-soaked material and the worms may actually exhibit symptoms identical to "protein poisoning." Or very faded white worms are found dead or dying in the bedding. Such bedding may not give a positive or strong reaction when tested for acid. The treatment for this is to turn the bedding with a fork and work it into smaller pieces if possible, thus releasing oxide gases being held by capillary attraction in the water. It should be exposed to sunlight if possible and a fan or other source of air movement will help to dry it quickly.

Use any kind of limestone or antibiotic sparingly in very wet beds. Most acid antidotes will form nitrogen in the presence of water and the worms must have oxygen to live. There is little danger of over use if bedding is properly damp instead of soaking wet.

The digestive organs of an earthworm are equipped to convert cellulose and starches to energy. Leaves and vegetable products of every kind are full of cellulose. Grain contains proteins and starches, some more than others. All the cereal grains, corn, oats, wheat, barley, millet, etc. are rich in starches and sugars and some, notably oats, wheat and corn have a high protein content. The shells, husk and chaff (bran) of all grains are particularly rich in convertible cellulose. Many of the natural vitamins, A, B, and C are also present in the brans.

Bran or chaff of all kinds, while the effects are not immediately apparent, can and should be fed often. The cellulose becomes available as tissue building food as soon as oxidation and bacteria break it down, or if it is ground to a fine flour the worms will eat it and benefit immediately to some extent.

Bran and chaff of all kinds may be converted by the worms more readily if it is soaked in a very mild solution of acetic acid before feeding. Ordinary cider vinegar contains 4% to 8% of acetic acid. It should be used by adding one tablespoon of vinegar to half a pint of water. This should be poured over one half gallon of bran and the bran immediately stirred or mixed so that the acid will be absorbed evenly. Acetic acid may be obtained from the druggist in any required strength and should be experimented with to some extent, using a strength of 5% or less until results are observed and established. Acetic acid will affect the production or fertility of capsules to some extent but will pay for the loss in additional size on premium worms.

Calcium carbonate will also counteract acetic acid if too much is used accidentally. Antibiotics are not effective in any case of too much chemical acid added to the beds. Calcium carbonate kills bacteria by destroying the acid which they have produced to digest their food. Antibiotics kill germs by direct attack on their chetin (skin or shell) thus reducing the amount of acid produced in the beds.

The sulfa drugs are well known and powerful bacteria killers, but are not to be used in the worm beds. They will not harm the worms to any extent but may be dangerous to pickers who are often susceptible to violent reaction when they come in contact with such drugs. Hands often become red and painful. DO NOT USE THE SULFA DRUGS IN WORM BEDS.

The antibiotics are usually applied with a large salt shaker or foliage duster, and are applied directly to the surface of bedding that should be wet. Allow the antibiotic to remain at least twelve hours after which the bed should be turned, dampened if necessary, treated again and allowed to stand a day or two before watering. Feed may be added over the antibiotic.

Usually two treatments of antibiotic of about one ounce to ten square feet of surface area suffices to cure the condition, although results may not be apparent for several days, as those worms affected less severly may continue to come to the surface and die.

Remember that the sour bed may have taken weeks or even months to develop, and a cure cannot be effected overnight. Three or four applications of calcium carbonate three days apart will usually cure even the most stubborn bed. There is no danger of using too much of any of the described materials as continued feeding and care will start bacterial action again immediately. We use about one pint of calcium carbonate to each ten square feet of bedding area per application.

Reference has been made here to some materials that are scarce or expensive in your area. In most cases satisfactory substitutes will be given or can be found. Think out your problem for yourself. The very best of written instructions are only an incentive and guide to independent thought and action. An instruction book is a tool as is a saw or hammer, learn to use it to shape the product to your needs.

I will repeat again in this book: Crushed limestone—calcium carbonate—is easy to find anywhere in the world. It is used in all poultry feeds, and is the material used to make the lines on baseball and football fields. Hundreds of people have written to say that it cannot be found in their areas...but it is plentiful and cheap everywhere.

FORCE GROWING (FATTENING) WORMS

The problem of "force feeding" or growing out large worms, making big ones out of little ones quickly and cheaply, is a tantalizing and serious problem for all worm growers.

As the season advances and fishing is at its best the weather grows hotter and may be drier. Your worm seems to develop slower and slower. The greater the demand the scarcer salable worms become. Usually the beds are teeming at this point with little fellows too small to sell.

The several solutions to this problem are all expensive or time consuming or both. The favorite method with the inexperienced is to "pour on the feed" and try to force the growth so sorely needed. This usually results in sour, acid infested beds; but not always. Sometimes it may actually work to some extent--but it is certain to lead to trouble eventually in many cases the trouble occurs so much later that the grower makes no connection between his actions and eventual results.

The two certain methods that will always work are those least used in actual practice.

The first, and best, takes more time, or labor, which in a large operation where help is paid by the hour may become very expensive: Build or have ready one or more spare beds, complete with aged or previously prepared bedding soaking wet but unfed.

Pick out older collared breeders from the crowded bed or beds and stock the new bed at the rate of only about 200 or 250 worms to the square foot of bedding surface. If African worms, only about 125 or 150 to the square foot. The number is not critical but the worms should have plenty of space. Feed them as soon as they have gone into the new bedding. Prompt feeding will often prevent the crawling which takes place sometimes when worms are transferred into a new bed.

The above method, as do all others, has its limitations and frustrations. No sooner do the worms find themselves in a new uncrowded environment than they start trying to produce enough young to fill it as full as possible in the shortest possible time.

Fortunately there is a time lapse between capsule production and hatching, and during this time the worms will grow a great deal, sometimes doubling and redoubling their size. The worms in their new bed should be fed heavily, as much as they will clean up overnight. Always feed such new beds on the surface so that feeding activity can be observed and the amount of feed controlled to some extent.

Worms in fattening beds should not be fed garbage and massive amounts of waste products. Laying mash is the best feed to use here. The addition of pulverised sewer sludge, 30% of sewer sludge, 70% of laying mash by volume, will often not only reduce the number and fertility of capsules but will enhance the growth of the worms to some extent. Laying mash or other types of feed here, as elsewhere in the worm operation, should be ground fine. The finer the feed the better job it will do, and some of our finest worms were produced on chick growing mash ground almost as fine as cake flour.

Usually if the worms being transferred are collared breeders, you can expect them to double their size in from seven to ten days. This is under perfect conditions with proper care and feeding. Their length may remain much the same, but they will gain the bulk that appeals to fishermen and people who are buying breeder worms.

One other good method that should be more widely used than it is, is described in a book titled "Larger Red Worms" by M.W. Holwager. It describes in considerable detail the use of sewer sludge as a fattening agent for "hybrid" and other red worms of Helodrilus, Eisenia and Allolobomorpha families, which comprise most of the worms grown commercially in the United States and Canada except the African nightcrawler. "Larger Red Worms" is available from the grower from whom you purchased this book or from Shields Publications,

Another common method to produce large worms is by "splitting out" the beds. To use this method you simply halve the material in one bed and place half in another empty one. Then use fresh bedding to replenish both beds. The main drawback to this method is that young worms and capsules are also transferred and the process of overcrowding again starts immediately.

In either of these methods feeding, both the kind of feed and method of using it, is very important. Best results are obtained with any kind of worm by surface feeding during hot weather. Surface feeding means placing the feed on the surface, or top, of the bedding. Ground grains and sewer sludge are spread by hand over the surface of the bed AFTER it has been heavily watered. Feed should be used sparingly at first then enough added each feeding until you are giving them all they will clean up. There should be no feed left on the bedding on the following morning.

Fattening beds should be fed every day and if the worms are particularly active, as they may be within a day or two of transferring, feed them twice a day. Feed each morning and evening, with the heaviest feeding at evening because they are much more active at night usually and will eat more than during daylight. We feed our own beds every day, winter and summer, although outside beds will have no activity during very cold weather. Ours are all inside during cool and cold weather. Sometimes during heavy winds or particularly dry weather the worms will feed much less than in normally still hot weather.

I repeat: Fattening beds should be watered and fed daily. Breeding beds may perhaps be fed only when the supply of feed in and on the bed is exhausted. The problem here is to know when the feed under the surface is gone as very heavily populated beds of all sizes of worms mixed will drag large quantities of feed beneath the surface. But fattening worms is no different than fattening any other kind of animal, and a constant supply of food is essential in order to avoid movement of the subject so far as possible.

Our own beds contain African worms and there are eight to sixteen large beds, depending on needs and season. We double the recommended population because of several factors, chief of which is experience. We water them heavily daily and feed morning and evenings with a finely ground 20% protein lay mash. The beds are built above ground, with no bottom except the soil. These beds, when heavily populated, will require 800 to 1000 pounds of feed per week. These are fattening beds.

The propogation beds, perhaps five times as much area as our fattening beds, require about the same amount of feed.

The feed found to be best and most consistently trouble free, winter and summer, is triple ground (very, very finely powdered) 20% lay mash. Growing mash is just as good. To the mash we add 10% by volume of dehydrated alfalfa leaf meal. The feed is scattered on the surface by hand and is never raked or watered in. Surface feeding of all beds also tends to keep the worms near the surface, which greatly facilitates picking.

Since the above was written, a new, more trouble free feed has been found by experiment. This is: 65% ground peanut hulls, 10% fine ground corn meal, 10% crushed limestone (by weight) and

15% dehydrated alfalfa meal. You may have some trouble, depending on where you live, finding peanut hulls, but if you get serious with your feed supplier, he will find and obtain them for you. If you cannot obtain ground peanut hulls, get crushed or whole and have them ground or run them through any kind of beater and they will be beaten enough to use. *(THIS FEED IS ABSOLUTELY TROUBLE FREE. IF YOU HAVE TROUBLE AFTER USING IT, LOOK FOR THE TROUBLE FROM ANOTHER SOURCE.)* Of course any food can be overfed.

Various methods of feeding are recommended by growers for different types of worms. All worms will work near the surface if the feed is there, and most worms will at some time during the day or night come out on the surface of damp beds. They migrate from one part of the bed to the other in this way, or they may just crawl about. It is surmised that they do this in order to replenish the body oxygen, which may be in short supply under the surface of heavily populated beds.

Bedding is seldom more than six inches deep for fattening African worms or more than ten inches deep for other types. All commercially produced types are fed and watered in the same fashion except that sewer sludge is not used for African nightcrawlers. A few people have worked out methods of feeding special materials and these methods may vary greatly from those given here. Our own trials prove that these special feeds could be used as well or better by our method--and with considerably less work.

FEED LIGHTLY at first for all types then add to the amount each day until some feed is left on the bed in the morning. Then cut back the amount so that it is just cleaned up every night. Weather changes may affect the feeding a good bit, especially if it turns cool or dry. If some feed is left on after several days of good feeding it means that the worms are probably as large as they will get on this move and are ready to pick either for sale or for removal to a third bed, which some growers do.

The factors that might cause worms to reduce their feeding activity are a hot dry wind, a sudden change in temperature, either hotter or colder. They will feed much better on warm damp still nights--and usually worms will eat about twice as much during the dark hours as during daylight hours. They may not feed at all during the day if they are not in deep shade or covered.

My own fattening beds are under shelter (sheet metal or plastic) and in deep shade, and I consider shade an essential factor in proper growth and fattening of worms. Roofed protection combined with a shade cover gives both cooling and perfect control of the growing area, as too much water over a long period of time causes crawling in almost all types of worms and even if the worms do not crawl out of the beds they will lose weight from movement back and forth in the bed when rain is beating on it.

It is possible by the method of picking and moving the older worms, to grow out five or six inch wigglers or manure worms by moving two or more times. You can estimate whether or not the enhanced demand and value will make it worthwhile for you. Many growers move worms, especially African nightcrawlers, up to four times in order to be able to advertise and sell "foot long" worms. Naturally these monstrosities command a premium price, and they are worth it to the fisherman who is convinced that the size of the fish is commensurate to the size and quality of the worm. I am a much more avid fisherman than I am worm grower, and I am not convinced that the larger worm catches the larger fish--although I must admit, in answer to hundreds of questioners: I seldom or never use worms for bait. I am a fly fisherman. A word here is to beware of those people who advertise "foot long" worms then ship "foot long types" that may be much less than half a foot long.

Under ideal growing conditions almost all types of worms, in common with their more distant relatives, will continue to increase in size as long as they live, and most of the common types have a life span of five years or more. Most commercial worms are taken out and sold before they are six months old, therefore it can be calculated that they are only about 1/5 of the potential size at a most conservative estimate. Of course they develop faster during their first three months of life than they ever do thereafter, except under very unusual conditions.

Recent experiments and observations by naturalists reveal that almost all the Eisenia types (manure worms, "hybrids," etc.) will grow to meet the contingencies of life-when well aged manure worms accidentally get into a bed of large African worms they will very shortly become as large as they must become in order to compete with the other worms for food. It is not unusual for one to grow to six inches long within two or three weeks time. This is sometimes called homomorphous development, and is common in many forms of life. Each form is peculiarly adapted to its environment, and many of nature's forms have the ability to adapt almost immediately to changes in environment, while others require thousands of years to change even slightly.

Earthworms are so adaptable to environment that, providing food is plentiful and moisture normal, they will grow to actually resemble to some extent in color, and to a great extent in form and size, any type of other earthworm with which they may be forced to live in close proximity. For this reason many growers have claimed to have developed crosses or hybrids, and they may actually think that they have done so. It may require dissection and microscopic examination of the subject in order to identify it for what it really is. A highly specialized segment of biology is taxonomy--the science of recognizing species.

Recognition and classification of any particular animal form may be a long drawn out process, even by an expert. Still another branch of the science of biology is that of "fixing" microscopically thin slices of tissue that can be studied by the taxonomist and compared with other slices of a known form. Thus only can true identification be accomplished. Any farmer can tell a cow from a horse or a hen from a pig, but very few worm growers can distinguish between the manure worm (Eisenia foetida) and the so-called "hybrid" which may be Helodrilus (Allolobomorpha) caliginosus. Yet there is a very distinct difference readily recognized by a taxonomist and which can be recognized by anyone when the two worms are found separately in nature. The trouble is that these two worms are found together in every bed that I have ever examined that had been established for six months or more. It is almost impossible to establish one "strain" or species of common earthworm and maintain it without other types somehow getting in. Wild worms have a positive genius for finding their way into places that are especially suited for them, and much effort is expended in making a worm bed especially suitable.

* * * * * * * *

SYNTHETIC MANURES

The procedures in this chapter should be followed closely. Read the directions as you proceed. Keep them before you at all times when preparing and using the feeds. Do not attempt to go from memory except after repeated performance of the processes involved. Measures in most cases are not arbitrary. They may be varied slightly without prejudice to the final outcome, but it is better to follow directions exactly, then you are sure.

Many people are disgusted by the odor and appearance of animal or fowl droppings in any of their forms. Such droppings are not available to many people while others find their use objectionable because of proximity to neighbors, or because of laws enacted to prevent unsanitary or careless use of them.

In the United States the word "manure" has come to be a common term for animal droppings while in most other English-speaking countries manure is a product made of green or cured vegetable matter composted in pits or chopped by mechanical means for enriching the soil.

The use of manure of almost all kinds is one of the most common and better methods of feeding worms and increasing productivity of the beds. All worms thrive on animal and fowl droppings, especially that of rabbits, cows, and horses. This includes northern native nightcrawlers and garden worms, which prefer lightly manured soil, wigglers of all types, brown nose worms, and the African nightcrawler or tropical worm. Without exception both domesticated and wild worms are more numerous where animal droppings are found naturally in pastures, feed lots, barn areas, rabbitries, poultry runs, and even pig sties.

The natural manure is attractive to worms because of the fact that vegetable material is found here concentrated into the most nourishing form. It has been predigested and ground to a fine powder. It has natural moisture retaining qualities and many enzymes and antibiotic products are inherent in it. It is not that it is manure as such that makes it attractive to worms, but that the left-over products of animal digestion are nature's finest source of worm food.

To make synthetic manure there is needed only some vegetable and/or grain products such as animals or fowl eat, and a couple of easily available items from the drug store. By vegetable product I mean ANY vegetable product---anything that can be ground or chopped finely enough may be used in either raw or refined form. Grass clippings, hay, waste vegetable products of any kind or form including roadside or fence line weeds, green leaves, corn chop (corn, husk, and cobs ground together), or any ground or crushed grain, cottonseed meal, peanut meal, hominy feed, rolled oats or barley, waste products from processing and feed factories and canneries, mill sweepings of all kinds, processed foods such as corn flakes and other prepared cereals, in other words, ANYTHING.

Synthetic manure is better if made from a mixture of ground grain and chopped fibrous vegetable material. Any kind of cured hay, finely chopped, is suitable, or field or garden trash with the exception of tomato, potato, fennel and other highly odorous growth, and it will not matter if a little of this gets in.

To prepare to manufacture synthetic manure you first obtain a solution of 5% hydrochloric acid. The druggist will make this up for about 50¢ a pint. You will also need some pepsin, also obtainable from the druggist in bottles containing 2, 4, 6 ounces or larger. Unless it is a large well stocked drug store they may have to order the pepsin. Your veterinary supply store, your druggist, or the veterinary Doctor will supply one of the mycins, there are many of them. The prefix is usually a trade name; Auro-mycin, Strepto-mycin, Galli-mycin, etc. Any poultry supply store has or will obtain terramycin in small 4 ounce bottles or can obtain it in larger quantities up to five pound tins. Pfizer Terramycin poultry formula anti-germ No. 77 is one full name.

If the mycins are too much trouble or too expensive use liberal quantities of crushed limestone or crushed oyster shell. If you bring the material to a boil over a slow fire the bacteria will be destroyed much more quickly and effectively than by any antibiotic or mineral.

Start with a chopping block of some kind. It can be a regular meat block, a section of board, or a block of wood of any kind. A cutter will be needed--axe, hatchet, cleaver, machete, even a heavy butcher knife. (It is assumed that no silage cutters or grinders are available to the average worm grower although some will have them.)

Cut your material up as fine as possible without doing a lot of heavy labor (the writer, a middle-aged man of 140 pounds, has cut a half ton of this material with a hand axe in less than a day of not too strenuous labor) until you have a wash tub heaping full (or a bucket, if you want to start on a smaller scale). Be careful with your chopper, of whatever kind. Many years of personal observation have not established that worms are particularly fond of human fingers.

There are various kinds of small compost and organic grinders, mulchers, and choppers available as advertised in farm and garden publications that will do an excellent job of cutting all types of mulch. A lawn mower of the rotary type may be used by building a low fence and enclosing a small area on three sides where the green material can be piled and chopped by running the rotary mower back and forth over it. The fence prevents scattering of the materials. There are also mulcher attachments available for other mowers, and these make satisfactory grinders for all green materials.

All vegetable material to be used in synthetic manure should be either green or cured green in the manner of hay, not dead standing or fallen material such as dead leaves or frost killed weeds. Such dead standing and fallen material may be used for bedding or as a bedding additive, and may be used in manure but it takes very much longer to process them and the vitamin content is much lower. All dried materials such as hay should be soaked overnight in plain water before being chopped, and even then it might be tough and hard to cut.

The second step is to prepare the hydrochloric by diluting it one part to nineteen parts of water. A regular measuring tablespoon, available at any grocery or ten cent store, holds half an ounce. One and a half tablespoons of 5% hydrochloric in a quart of water makes a solution of ½ of 1 percent hydrochloric acid. A little weaker or stronger will make no appreciable difference in the final product, but it is better to be exact and be sure.

To each quart of hydrochloric mixture add one half teaspoon of pepsin crystals. Pepsin in crystal form is most common and easiest to obtain. If only powder is available use ¾ of the measure given above. If liquid, use one teaspoonful.

Then, using a tub, bucket, or any convenient receptacle--half an oil barrel is easily obtained and ideal for this as well as other things which will not leak, tamp or trample about six inches of the chopped material into the bottom of your container.

Cover the bottom layer of vegetable material with a quarter

inch of ground grain. This can be anything--corn meal, snapped corn, lay mash, dairy feed, chick start, rice flour, barley flour, grits, rolled or pulverised oats, barley or other grain. Be sure that any grain used is ground or crushed, the finer the better, but if it is flour, fine, only put half as much, or ¼ inch instead of ½ inch. A thin covering, ¼ inch or less, of cottonseed meal, peanut meal, or soybean meal may be used in lieu of the grain. Dampen the grain or meal with plain water and a light sprinkling of the hydrochloric solution.

Continue these layers of vegetable fiber and grain until your receptacle is about half full. Pour the remaining hydrocloric-pepsin solution over the top. Work the material up and down by pressing and releasing to be sure that it contains plenty of water. It should be very wet and when the hand is pressed into it water should show in the depression. It is impossible to give exact amounts of water due to the varying conditions in different climates and the fact that all growing things contain more or less water depending on the season.

When the several layers are made you will have used from 2 quarts to a gallon of the solution in a number 3 wash tub, depending on the amount of moisture (again!) in the material.

After the layers are placed and the mixture is found to be almost flooded with water, soaking wet, use a package of dry yeast (a cake of the refrigerated kind will do here) and dissolve it in a pint of warm water. Do not use boiling water or cold water. It will kill the yeast germs. Pour this mixture over the material then place a couple of boards, weighted with a large stone or other heavy object, over the top of it. The boards should be cut short enough so that they will fit down inside the container and press directly on the manure. This is to prevent the bubbling action of digestion from swelling the material over the top of the container.

As fermentation takes place bubbles will form in the solution and the liquid will increase in volume. The grains and other material will absorb moisture and expand, causing the whole mass to increase in volume to almost twice its original size.

Yeast is suggested here because it is easily available. One of the common septic tank activators would be much better. Activo or Northel activators are good and have proven themselves in use. Northel Reactivator may be obtained from Northel Distributor, Box 1103, Minneapolis 40, Minnesota. Use a level teaspoon of activator but be sure that it is a bacterial concentrate activator, and not a lye or caustic.

NEVER ATTEMPT TO MAKE ANY KIND OF SYNTHETIC MANURE in a tightly closed or sealed container. THE EXPANSION DESCRIBED ABOVE WILL CAUSE A TREMENDOUS PRESSURE TO FORM, AND ANY ATTEMPT TO MAKE THE MATERIAL IN A CLOSED OR SEALED CONTAINER MAY CAUSE A SERIOUS AND DANGEROUS EXPLOSION.

The synthetic manure mixture should always remain in a warm place, preferably under shelter where rain cannot get in it. It will not work properly in a temperature lower than about 60 degrees. If it is made in outdoor pits, cover with a loose sheet of tin or other material to keep out rain water.

The mixture will begin to ferment in a few hours, depending on the temperature and foam will form on top to some extent. The working of fermentation will further mix the materials and grain, and after about two days in 70 to 85 degree temperature, or less than two days at 85 degrees or over the materials should be well broken down. The top will not show it, and there will be stalks and fibers left, but the cellulose, vegetable proteases, minerals, and vitamins, will be suspended in the liquid, of which there will appear to be too much.

At this point the mycin is used. A number 3 wash tub will need about 3 tablespoonfuls. In very hot 90 degree and over temperatures use 5 tablespoonfuls to kill the bacteria and stop the working of the material. The mycin should be sprinkled on top and then stirred in. The fermentation will subside but probably will not stop altogether.

In the absence of one of the mycin antibiotics, cooking or boiling will stop fermentation completely but temporarily, as it will start again within a short time after it is allowed to cool. In fact, any treatment is more or less temporary as the effects become less and less apparent and new bacterial growth starts. For this reason you should only make as much as will be used immediately. It cannot be satisfactorily stored without a good bit of trouble.

The mixture will not be too odorous at this stage. In fact, it may be made in any building without offense, but when fermentation is stopped and decay bacteria move in, then it will begin to have an unpleasant odor. This is another reason for making it a little at a time.

As soon as fermentation is arrested, the material is ready to be fed to your worms. It should be further prepared at this point by adding one cup of calcium carbonate (crushed limestone) to each half gallon of synthetic manure. Mix this in well. In fact the manure may be fed without either boiling or the use of mycins if two cups of calcium carbonate are added to each half gallon of manure.

This is highly concentrated food, and must be used carefully to avoid overfeeding the worms. Use it by making a small trench down the length of the bin or bed, a trench on each side of very wide beds, and pouring it lightly in the trench. Cover lightly with bedding to prevent drying and hardening. The worms will probably avoid it the first day or two but then they will be found working around the edges of it and soon, if the bed is not too crowded, you will have some of the largest worms it is possible to grow. This mix is good for all types of commercial worms except the northern native nightcrawler, or dew worm. It is especially effective with the African nightcrawler and all members of the "wiggler family."

REMEMBER!!! FEED LIGHTLY!!!

The foregoing is for making "synthetic" manure from hay or pasture fed cows, and the final result will contain all of the desirable elements found in such manure, and it will be much cleaner, less odorous, will not attract flies if fed lightly and covered lightly, and will not be seeded with undesirable "wild" worms that are often, in fact, almost always, found in feild manure, even stall manure to some extent.

"Grain fed cow" manure is easier, if more expensive, to prepare, and must be fed even more carefully than that made from vegetables or fibers and grain. It is very rich in protein and carbohydrates, and can be a source of trouble if overfeeding occurs.

Two gallons of lay mash, rolled oats, grits, corn meal, snapped corn, rolled or crimped or crushed oats or barley, crushed corn, or any ground grain or seed mixture such as are used for feeding poultry or stock including cracked bird seed should be used here. Or substitute one half gallon of cottonseed or peanut meal to 1½ gallons of grain. A mixture of several of the grains mentioned is generally more to be desired than any single one.

Put the mixture into a five gallon container or any size wash tub, mix plain water with it until it becomes a soupy consistency then add one quarter (¼) pint of 5% hydrochloric acid, one teaspoon of pepsin crystals or liquid, and one package of dried or one cake of refrigerated yeast. Again, Activo or Northel Reactivator will work better, but only add one teaspoon of it. Be sure you do not use one of the detergent or alkali septic tank cleaners, but bacterial activator only.

The yeast or activator should be mixed in a small amount of warm water and mixed in after the pepsin and hydrochloric. Stir the whole thing well and cover with a cloth or screen and allow to stand in a warm place, not below 65 degrees. The warmer (up to about 110 degrees) the better and the quicker the feed will be prepared.

The grain mixture will harden on top and the cake thus formed will swell over the top of the container if it is not stirred at least twice every day, more often is better.

After about three days, perhaps sooner if the storage place is very warm, the mixture will "calico" or begin to have streaks of clear water in it. If it is too thick the "calico" will not be apparent. Warm water can be added if necessary at any point to thin the mixture. Three days should be enough time to properly prepare the manure for feeding.

Snapped corn and other grains will gain at least fifty percent efficiency as a worm feed when treated as above before feeding. As soon as the mixture has stopped working to some extent, and the grain particles begin to settle to the bottom instead of floating, treat with two ounces of poultry grade Aureomycin, Terramycin, or one of the other mycin antibiotics. Simple boiling will kill the bacteria more effectively and quickly, and will be much cheaper. About two cups of calcium carbonate per gallon of feed should be mixed in as soon as it is treated with mycin or boiled--boiling means bringing to a slow boil then removing from the fire immediately as prolonged cooking will destroy essential vitamins. The treatment of calcium carbonate will prevent refermentation for a short time.

The grain manure is best fed by pulling the bedding in the bin up into a hill in the center, thus forming a ridge or hill the length of the bed Pour the grain manure lightly along the top of this ridge allowing the juice to run into the bedding. When the manure begins to dry or form a crust, sprinkle lightly with calcium carbonate, about one pint to ten feet of bedding, then sprinkle lightly with water. The manure will be kept damp on the underside from capillary attraction from the wet bedding.

These mixtures are wonderful additives to either breeding or fattening beds, and the addition of a small amount of riboflavin (also called vitamin B2 or G) will increase growth and size of your worms in fattening beds but may slow reproduction in breeding beds--fat worms are not good breeders.

In all of the above mixtures don't overlook the possibility of using any of the things that may be available to you at no cost. Everything from flue bran (a by-product of cottonseed processing) to waste fruit skins, sugar beet pulp, silage of any or every kind, at the risk of repetition--any vegetable or grain product used by men or animals.

****

FEEDING THE BEDS

There are almost as many different methods of feeding as there are people doing the feeding. In the foregoing chapter we described the hard way of accomplishing a simple thing--causing feed to sour under controlled conditions. This can be done in the beds just as well and can be controlled to a greater extent in the beds than it can in a container.

Many people want to know the hard as well as the easy way of doing things. Some people even prefer to do things the hard way under the impression that hard work necessarily makes for success. I don't happen to feel this way about it and always look for the easy way to do whatever must be done.

During the hot summer months and even, to some extent during spring and fall months when the climate is still warm, you can prepare "synthetic" manure right in the worm beds without resorting to the expedient of allowing it to ferment in a container.

The procedure is the same, with the same materials involved, in the same proportions. The only difference is that when the materials are ready and the mixture made you pull up the beds into a "hump," or ridge, in the center of the bin. Then use a rake or other tool and flatten the top slightly. Using the same rake then cut a shallow trench down the center of the flattened area at the top of the pile of bedding.

The wet, but unsoured material is then poured into the trench, and the excess bedding on the side is raked over the material lightly to cover it. This light cover prevents flies and other insect pests as well as excess evaporation of moisture. The trench is used with any and all kinds of manures, and is more efficient than other ways of using the real manure or animal droppings.

Within a day of placing the manure in the trench it will begin to heat and the worms will leave it strictly alone but within another day or so the edge will cool and the worms will be found there in great numbers. Mostly the very large worms will leave the synthetic manures strictly alone, going to it only for short periods maybe during the cool of the night. The small worms, however, will find their way to the manure and will usually, except in the hottest weather, burrow under it and feed on the juices soaking into the bedding underneath, as well as crowding around the surface edges.

All beds with feedings of synthetic manure should also be fed on the surface in the normal way. This is by spreading or scattering small amounts of lay mash or other type feed on the surface after the bed has been watered. This is particularly effective in separating the large worms from the small.

As stated above, large worms will usually avoid heavy concentrations of feed after they have reached their full growth. And they attempt to avoid, as much as possible, disturbance by young worms. This makes them seek the sides of the humped beds. By feeding the sides, then, you can attract the large worms to the surface where the pickers can take them out without disturbing the center row of feed, which should not be disturbed or torn apart in any case.

This method of separation works for all kinds of worms in any size bed. In hundreds of trials it has never failed to work except in very cool or cold weather when beds should be leveled and fed in two or more trenches along the sides.

It is well to remember that all adult worms of all types attempt to avoid concentrations of young. You can take advantage of this by so arranging your feed that the young worms will be concentrated in one area. This will naturally drive the older larger worms into more or less concentration in another area where they may be easier to find or separate into smaller numbers.

In cold weather the young worms will normally be found at a lower depth than will the larger worms. Of course the separation at all times will depend on food supply. The larger older worms will readily mix with the young in order to compete for food--on the other hand the younger worms will, in cool weather, mix with the larger worms if they must in order to find food. And of course sometimes under any given conditions the worms will be found mixed, or roped into a mass. This only proves that there are no hard and fast rules to go by when dealing with live creatures of any kind.

The earthworm may not have a well developed brain or a "psychology" as such, but the worms do have a well developed instinct. This instinct tells them what is good and what is bad, to some extent, and usually when you observe some strange action on the part of large numbers of worms in the beds it is time to start looking for trouble. Though what is strange to you may be perfectly normal behavior for the worm under the conditions that exist.

Sometimes when beds are thinly populated and an abnormal amount of propogation is desired, or when a beginner is starting with the fewest possible number of breeders, there is a method that we use to start heavy reproduction in any bed in which it is desired.

We remove the bedding from the center of a bed, the full length of it, to a width of about one foot and to the bottom, or about eight inches if it happens to be deeper than that, which mine seldom are. The removed bedding is piled to the sides in order only to clear the center space.

We then set two boards, one by fours, or one by sixes, in the cleared space and about six or eight inches apart, and extending the full length of the bed or filling at least one section of it as long as the boards will reach, after which they are moved on to the next section to be fed. It is best if they are very long or if two boards are fastened together to make one long board for this purpose.

Then we make up a mixture of lay mash and dehydrated alfalfa leaf meal by using 70% lay mash and 30% leaf meal. To this is added an equal amount of peat moss (by volume, not by weight), then the whole is wetted until it becomes a thick soupy mass, about as thick as a thin corn bread batter.

This mixture is then placed between the boards in the center of the worm bed to within an inch or so of the surface level of the surrounding bedding. We remove the boards, rake the previously removed bedding back over the core of feed and peat moss and there you have one of the finest propogation beds it is possible to prepare.

Such beds are particularly effective in winter, early spring, and late fall, and are an effective and efficient method of "heat feeding" for any cool season. The heat from the material will radiate into the bedding in all directions and the worms will be attracted to it. There is plenty of room, or should be, for the worms to move away from the heat if it becomes excessive or harmful. This method should not be used in very narrow beds, or the width of the core of feed should be narrowed accordingly. In a short time it will cool enough that the worms will go to it and start feeding, and producing capsules at a rate sometimes four or more times heavier than usual.

Natural cow or horse manure makes an excellent material for these "heat cores" and when used should be fresh green manure that will heat over and over when it is watered. The grain and peat mixture will heat each time it is watered, and this may go on for several weeks, depending on the climate, thus furnishing a continuous supply of heat to the beds.

Beds treated with a core of feed or manure should still be fed small amounts on the surface, as some of the larger worms will avoid the core because of the probable concentration of small worms there. If the worms do not feed on the surface then surface feeding may be discontinued, but careful watch should be kept, and surface feeding resumed when it is needed. This will be indicated by many large worms being found along the sides of the bed.

One thing that has probably caused the loss of more worms and the discouragement of more worm growers than any other one thing is the problem of winter and early spring feeding.

It is well first to reread the first two chapters of this book and refresh your memory of what goes on in a worm bed, no matter what kind of worms or bedding is being used. We deliberately avoid the discussion of beddings here. The type of bedding is only important as it may be more or less available for use or expensive for you to use. The methods of feeding given may be used with any bedding that you may happen to be using or have at hand to use. You must exercise your judgement in the amounts of feed to use as the requirement may be reflected by the amount of natural feed in the bedding you use. Most beddings have some natural feed, but this is soon exhausted in a heavily populated worm bed. In fact, most beddings heat, and the process of heating destroys most of the vitamins and converts some of the minerals to forms indigestible to any life form. Thus the necessity for adding fresh feeds at suitable intervals.

Winter and cool weather feeding presents several problems that are not apparent at the time you are placing the feed, and may not become apparent until several weeks or even months have elapsed.

During cool weather the metabolism of all cold blooded forms is reduced. Many of these forms are completely demobilized at freezing (32 degrees) temperature while some can and do carry on some activity at 32 degrees or even a little lower. Some forms are completely inactive at higher temperatures, notably the African worm. Depending on age or physical condition they may actually die at temperatures as high as 42 degrees while the younger more vigorous worms may withstand and survive a temperature as low as 32 degrees.

Because of the reduced temperature both the activity of the worms and the bacteria with which they live are reduced. The worms may continue to be active in the bedding to some extent therefore they continue to bury the feed you put on the bed, and may actually ingest it without digesting it. This feed then accumulates, and can easily fool the grower because it does oxidize and becomes the same color as the bedding that surrounds it. Thus the feed accumulates without being destroyed either by the worms or by normal bacterial action.

Then warm weather or several successive warm or hot days occur. The bacteria, and the worms, become active. The worms stir the bedding (and maybe the grower helps because the bedding may appear packed at this point, and the bacteria flourish, and the grower pours on the feed because spring has sprung and fishing season will soon be here, with its worm fishermen and dealers yelling for big fat worms.

The trouble occurs, or begins to occur, at this point. As the temperature increases so does the bacterial action and soon the bed is sour and the worms begin to squirm out on top and die. It is now too late to do anything about it. The worms have already developed the "disease" (it is not actually a disease as humans know it), and will continue to die for some time, possibly several weeks, no matter what type of treatment is begun at this point. But those worms least affected and the younger worms will recover and become good salable bait.

The quickest and safest thing to use at this point is one of the antibiotics. Terramycin or Gallimycin will destroy the bacteria immediately on contact if you use about twice the normal dose. Twice the normal dose would be two ounces to ten square feet of bedding surface. And use two treatments in quick succession.

Water the bed good, unless it is already too wet from melting ice, frost, or snow, then give the first massive (double) treatment. Wait three full days then turn and water the bedding (unless it is still sufficiently damp) and give the second massive treatment. Wait about four or five days then treat with a normal treatment. During the series of applications of antibiotic continue to feed but reduce the amount to a mere sprinkling. This will prevent crawling of the unhappy worms that survived. Also continue to water the beds lightly in order to maintain normal moisture as well as to leach the antibiotic into the bedding.

During the prescribed treatments we assume that a free flow of fresh air is allowed over the beds. The worms need plenty of fresh air at this point but it is possible that cold weather has returned and you may lose more worms by freezing than you would by not allowing the cold air in. The antibiotics may be used safely in a closed tight building as outside, in fact may be more effective inside as they are not washed away by unexpected rains. But sour beds are more quickly cured by a combination of treatment and fresh air. Bright sunlight is a good thing to have at this point if possible, it is not a necessity, however.

The calcium carbonate treatment may also be used in a situation of this kind. Here again we double the amount, sprinkle it on heavily, allow it to remain 24 hours, then turn the beds. Water the bed lightly then treat heavily again. After four or five days, treat heavily again. Also continue to feed and water during the calcium treatment but feed lightly here, too.

Of course the real sovereign remedy for the condition described above is to prevent it by drastically reducing the amount of feed when cold weather comes. The reduction may be accomplished by more than one method. You may know how much feed a given area has been using. If you do you can reduce the amount accordingly. Or results may be checked by taking some of the bedding into a warm (72 degrees or higher) room and "culturing" it there to see what is happening. After it is cultured four days in a warm place, test it with a soil test kit. The result will indicate the action you should take. If too acid, reduce the feed. If normal, reduce the feed a bit anyway, as a precaution. Worms, being more or less completely inactive during cold weather, will only shrink a bit if underfed. They will quickly regain their size when the weather warms and feeding conditions return to normal.

We have referred to the condition that kills the worms in an acid or sour bed as a disease. Actually it is not a disease that enters the blood stream or attacks an organ or nervous system as in animals. There are several things that may happen.

In one, the most common, the bacterial action is so rapid that the calciferous glands of the worm can not release enough calcium (calcium carbonate) to counteract it. The crop may swell and burst from continued fermentation after the food has entered or small bits of food may swell after further travel into the digestive system. In this case a rupture of the intestine may occur, and swelling at that point will appear as a "knot." The gizzard may be blocked and either swell or burst or the blockage may cause the worm to die of starvation. The most common cause of actual death in the case of "protein poisoning" (another name for the symptoms described above) is an infection of the cylia (kidneys). This makes it impossible for the worms to dispose of waste liquids, therefore the skin will rupture in some places while remaining swollen and unnaturally red in other places.

We have covered some conditions here that were previously covered but this information is so important that we believe it will stand, indeed require, repetition. Many well established worm farms have been completely destroyed because the grower could not perceive the relation between too heavy winter feeding and worms that die in the spring because of sour bedding.

****

FEED GROWN IN THE BEDS

Recently I experimented with methods of "growing feed" right in the earthworm bed. This proved not only practical but very helpful in growing larger, healthier worms. It also saved a good bit of the money formerly paid for grain feeds.

This may also be called a method of "growing manure" in the beds, far away from the animals usually associated with the product.

Webster's Unabridged Dictionary defines MANURE: n. Any matter which fertilizes land, as the refuse of stables and barnyards, guano, ashes, and every kind of animal and vegetable substance applied to land or capable of furnishing nutriments to plants.

Unfortunately Webster was not a worm grower or he would have included worms in his *"nutriments"* description.

So much for Webster. Originally manure meant the green organic matter of standing crops that was cut or plowed into the land; and the word is still so used in Britain and some of the Dominions, as in Canada. The history of this word led to wonder, the wonder to research, ther esearch to some of the usage of the English language.

And so the train of thought. The worms never make use of all the nutriments in the soil, if they did there would be nothing left for plants after the worms are through with it. Why are worm castings so rich in plant foods and nutriments? Obviously it is because there are great quantities of minerals, vitamins, vegetable salts, chemicals and unconverted proteins that the worm's digestive system is not capable of absorbing.

So why not convert this conglomeration of life-giving minerals and vitamins to a form that the worm can use? Plants can use them, and worms can eat plants after the plants are decayed--so why keep hauling in vegetable matter and adding it to other useless vegetable matter?

We went to work on this, then, from a worm grower's point of view. What can a worm best eat? Broken, crushed, or ground grain. That is one. Decaying vegetable matter. That is another. What do worms eat in their natural state? Certainly not crushed or ground grain. Not fruit, as the worm's natural habitat would produce little or no fruit to fall and decay. We hied ourselves to the nearest animal husbandry library and started doing some research on another train of thought.

What product, in its growing state, while very young, and that is very common, produces the highest protein and mineral supplement? Growing oats were found to be outstanding in this respect. Oat sprouts about six inches high, together with the roots and remainder of the seed itself, which adheres to the root system, are one of the richest common sources of protein. Protein that is immediately convertible to the use of animals of all kinds, including worms.

We first experimented with sowing oats thinly over the entire bed, standing by until they were about six inches high then digging them under. We had sown far too many, and those with even a leaf blade exposed did not rot, or even die, they kept growing. This gave us a lot of trouble. We almost never got the stuff killed off so we could even pick the worms much less so that the decaying matter would furnish food for the worms.

Remember that this was inside, in large beds, in a 50X50 foot building, which was heated.

Then we sat down to give the subject further consideration...We continued by finally sowing a thin row of oats down the center of the bed. They don't have to be covered, the worms take care of that.

The oat seeds will sprout in one or two days. If the beds are warm as normal beds are supposed to be for indoor propagation, with plenty of feed and water, the oats will grow to six inches high in about five or six days, even less during the real hot summer weather under shade or inside.

When the oats are about five inches high use a gardener's shears and clip the tops off close to the ground. The tops should be allowed to fall right in the bed. Let the tops wilt for one day where they have fallen then dig or turn the bed with a fork or potato rake, mixing the dead oat leaves with the bedding. They will rot there and are a wonderful substitute for nature's own high quality, high protein feed, manure--it is in fact manure, and will accomplish, so far as growth of worms and reproductive activity is concerned, as much or more than green stall or stable manure in the same quantity. And without any of the drawbacks of animal droppings. The flies and other pests of nature will never even know it is there--only you and the worms will benefit from the application of fresh manure.

We used clipped oats first quite by accident. These are available at any feed or seed store and I found them, in the long run, to be best, although ordinary feed oats will do the job about as well although they do not sprout so quickly. Be sure that you do not get any of the treated seed oats that might have a high percentage of insecticide or fungus preventive chemicals in them.

This method of growing manure works quite well in any bedding that we have tried it in, including the soilless bedding made from our formulas. It will in fact work well in any bedding, including animal manure.

Bed grown manure can also be made from many other plants, including corn, which is considered next to oats in nutriment. But of all that we tried, lawn grass and oats grew quick and decayed quicker, thus growing and decaying to clear the beds so that they might be worked.

The growing of green materials in beds can be handled in any bed on any scale from the smallest to the largest earthworm operation. A small sickle such as is used to cut high or tough grass in hedgrows and fields can be used to cut the growth in large or extensive beds. It also makes an excellent winter cover if it is allowed to grow to eight or ten inches then cut and allowed to fall on top of the beds. Several crops can be planted over the entire surface of the beds during the fall where each succeeding crop can be cut. The accumulation of material thus grown will heat to some extent and keep the worms in good condition for a longer period than usually is possible without cover.

One of the best winter covers I ever devised for outside beds was a thick growth of lawn grass from thickly sown seeds. The standing grass, not more than three or four inches high, was covered with a layer of burlap which was in turn covered with tarred roofing felt. This left a three or four inch dead air space above the bedding that was warmed by residual heat from the ground and the bedding. The worms can be fed and watered at will and when the opportunity affords by simply removing the tar paper and burlap, after which it can be replaced. The cover prevents the grass from growing too high.

Don't fail to give this method a trial, no matter what size operation you have. We produced some of the largest average worms we have ever produced, and on a smaller amount of surface feed, with oats grown right in the beds.

Green feed can be grown in beds even during the heavy picking periods of summer and cooler climate of spring and fall. The tender blades and stalks dissolve quickly into worm food after they are cut or separated from the roots. During the hot summer where beds are watered daily, the oats will sprout, grow, be cut and decay in six or seven days. The bed can then be picked and the green feed process started again. Our own experiments in both inside (during cold weather but in a heated building) and outside beds prove that up to 25% of your feed can be grown right in the worm bed. The worms can be fed ground grain or other feeds while the green feed is growing. Nothing changes except the amount of the feed bill and possibly the cycle of digging the beds and picking the worms.

In our own operation where we feed more than a ton of laying mash a week during hot summer weather a saving of 25% can be an important factor in our profit and loss statement.

FOOTNOTE: In relation to this type of bed-grown feed, we have recently been introduced to a plant new to us, Symphatum Peregrinum Ledeb, or Comfrey. This plant produces a leaf somewhat resembling dock and grows only from root stock. Dried samples of this leaf have been reported as high as 35% protein, and it is very prolific in all climates. Comfrey does not winter kill (the leaves turn brown, the root sprouts again every spring) and is very hardy, growing well in either dry or wet climates, cool or hot areas.

Our first trial of Comfrey was by growing it in a prepared area and feeding the cut green leaves to worms by burying the whole leaf in the worm beds. Our next trial was to cut some of the roots into small pieces and plant them right in the beds with the worms. During hot weather the Comfrey grows to a four inch leaf in about five days when buried in a worm bed. The roots are disturbed when the pickers go through the bedding for the worms but this does not seem to affect the growing at all. Even green growing out of the root will continue green, and will continue to grow. If they should be covered and die the worms will eat them and new shoots start immediately.

The main problem is that the roots grow continually larger and eventually must be taken out or cut into smaller pieces, leaving some to continue to grow in the beds. Pickers sometimes have a habit of tossing hard pieces and roots found in the beds away, this might also affect the production.

Green leaves of Comfrey, when buried just under the surface of the bedding and allowed to decay resulted in a very high rate of propagation, probably three or four times as many young were produced in this bed as in other beds without the Comfrey leaves. At least twice as many young were produced in this bed as were produced in beds fed a diet of fresh green manure. There was no effort in this test to compare weights of materials or control the propagation rate so it is not conclusive by any means.

Worms grow very fast on Comfrey leaves but the increased

breeding activity makes it a liability in beds set aside for fattening worms. Beds become very crowded quickly. We would recommend the Comfrey feed to all growers, especially those who produce worms for organic soil culture and where size is not so important as quantity.

This is not the ordinary American Comfrey, Symphatum officianale, but a specially grown plant whose root stock was imported from Asia, where both the leaf and the root are popular as food. It is said to produce as high as 36 tons of dry forage per acre, and is excellent for rabbits as well as worms and livestock of all kinds.

We have not had time, since acquiring the Comfrey root stock, to conduct any exhaustive tests but first indications are that it will be excellent as a bed grown feed as well as being excellently adapted to gardens and fields where it can be produced in quantity for feeding all types of domestic animals. It would be a wonderful fence row crop as it is tender and easy to cut with a hand sickle.

****

COMMERCIAL FEEDS

The variety of commercial feeds available to worm growers is exceeded only by the bewilderment engendered when said grower decides to feed his worms. There are mashes, chops, flours, pellets, liquids, kibbles, syrups, powders and heaven only knows what else because the worm grower has been told that what is good for other animals is also good for his worms. There are only a few domestic animals for which feed is manufactured. But the complexities of feeding those few animals can be realized only by considering the fact that more than six hundred different types, grinds, and mixtures of feed are offered for cows alone. Multiply this by horses, dogs, goats, sheep, hogs, chickens, turkeys, geese, ducks, rabbits and several others that I have left out and the total is truly staggering.

Only one manufacturer, so far as I know (and I have not researched the question at all, there may be others) manufactures a feed specifically to meet the needs of the worm grower. There are a few types of feed offered at different times by worm growers described as "miracle" feeds, or "secret" giant worm grower, and even an "imported" worm feed. The "import" can be discounted entirely and without further consideration because nowhere on earth are produced the variety and quantity of feed of all kinds, both human and animal, as are produced in North America.

The "secret" feeds and "mineral rich" feeds are to be considered only as curiosities. Transportation costs would destroy any added value that such feeds might possibly have. Whatever is available to the manufacturer of the "secret" feed is also available to you. "Vitamin enriched" feeds and "minerals added" feeds are available at your nearest poultry or animal feed supply house.

Practically every mixed feed of every kind that you may purchase has added vitamins and minerals for the benefit of the animal or fowl that the feed was intended for. These vitamins and minerals are also good for worms.

Any vitamin or mineral that is used in the feeding of any animal can be purchased by you at your nearest feed supply house. These come as supplements or additives, usually, and can be had in small handy packages or in bulk packages up to 100 pounds. People are particularly susceptible to vitamin claims. It is claimed that this or that vitamin is likely needed in not only your food but in the food that you feed your animals. Or it is claimed that this or that vitamin is lost in processing, or that mineral defficiencies arise from the loss of needed metals in processed foods.

The loss of vitamins and minerals is more likely to occur in human food than in any of the food intended for animals. Some processing systems bleach or cook foods excessively. This might result in the loss of some vitamin or mineral. Most animal feeds are not cooked or bleached or otherwise heavily altered in manufacturing processes, therefore are less likely to lose any of the vitamins or minerals that are necessary to healthy life. The bran, or outer coat, of practically all grains prepared for human use is separated from the finished product. These brans are rich in the B complex vitamins and fortunately can be purchased cheaply for use in worm beds, where they are very beneficial.

Vitamins are found in more or less quantity in various types of food as nature provides it. The present-day methods of vitamin production are, however, by chemical synthesis. In other words, they can be manufactured much cheaper than they can be extracted from the fruit or vegetable in which they are found naturally.

Those few that are not synthetized are very expensive or not available to the average user.

Minerals are processed from natural ores, for the most part, and are oxidized or otherwise chemically treated to permit ready assimilation by the bloood and tissue of animals.

Small town dealers may be poorly supplied with the vitamin or mineral supplements you may want to add to your worm feed. Some of them are not even very well acquainted with the problems encountered in feeding animals of any kind. In this case I suggest that you contact one of the salesmen who travels for a large organization such as Purina Mills, or X cel Feeds or whoever may represent the feed manufacturing interests in your area. Your county agent will have information. All of these people are familiar with feeds of all kinds, and many of them are animal husbandry or zoological majors, people who have devoted years of study to the problems of manufacture, distribution, and use of feeds of all kinds.

One of the most common of the prepared feeds is cottonseed meal. Cottonseed meal or cottonseed cake is used extensively throughout the south and to some extent in other areas. This is a fair product but not one of the best, by any means. It has a low price tag compared to other feeds, especially in the south, which is a recommendation in itself.

Cottonseed meal is oily and the main part of the food content is locked in this oil. Some of the vitamins are released freely but the protein, which is a necessity for proper and good worm growth, is suspended in the oil and only an acid action of some kind can release it. The worm cannot supply the acid from its own system so here we must depend on bacterial action. Sometimes there is a time lapse of six weeks or more between the feeding of cottonseed meal and its proper assimilation by the worms. The time element is controlled entirely by temperature, which is one of the reasons that southern growers are more often and severely plagued by attacks of "protein poisoning" than the average northern or western grower.

Here is the common example of "pouring on the feed to make em grow" during the fall and winter then finding all the large worms and sometimes the small ones too, squirming to the top of the bedding and dying by the thousands. Or traveling to the bottom of the bed and refusing to come up to eat or for anything else. Cottonseed meal oxidizes rapidly, that is, it turns dark on exposure to the slight acid of even a well cared for bed, therefore it is not readily seen, even when a large quantity of it may be already mixed into the worm bedding. And before it has had some bacterail action brought on by hot weather it will not give an acid reaction unless first "cultured" in a warm room for several days.

One bed that I examined in south Georgia proved, upon analysis, to contain over 20% of pure cottonseed meal. This had been added during the cool winter, had "disappeared" to all intents and purposes so the grower had added more, and more. Naturally his worms were dying. The mass of rotting cottonseed meal was enough to kill them without the acid incidental to its souring and fermentation.

Cottonseed meal is a good feed. Use it during warm or hot weather and use it carefully. Do not overfeed it--or any other type of feed, for that matter. It can be spread on as a meal or fed in trenches or can be mixed with water and fed in "dabs" or pats scattered around over the bed where the worms will congregate around it or can leave it if it heats or sours. It would be my recommendation to never feed cottonseed meal, or any of the oil seed feeds to worms bedded inside buildings. Use it only outside during summer. It is good in "cores."

Snapped corn, or corn chop (corn cobs, husks and grain all ground together into a coarse feed) is predominantly southern feed, although it is available almost everywhere, and a potential trouble maker in your worm beds. Corn chop consists of find and coarse material mixed, with the coarse material comprising much more than half the volume.

When introduced as feed in a worm bed, corn chop is quickly attacked by both worms and bacteria. The grain part is usually fine enough that the worms can ingest a large proportion of it and digest it to some extent. The cob part may be small enough to ingest but the worm cannot digest it because it is too tough to be ground up in the gizzard, it is then passed into the bedding much the same as before it was eaten. The husk part of corn chop consists of strawlike particles that the worm, and even bacteria, has trouble breaking up.

The parts of corn chop then that are not immediately assimilible by the worm accumulate in the bedding and when spring and warm weather come you are in trouble for the same reasons given above. In this instance even worse, because the corn product contains a large percentage of sugar and the sugar is converted to alcohol by the bacteria--and worms can withstand less of alcohol-type acids than any other of the several types of acid they have to contend with.

Corn chop makes a wonderful mixture for "core feeding" or heat feeding, as described in another chapter, but large quantities of it should never be mixed directly with the bedding or piled on top in quantity unless you know exactly what you are doing and have done it before. Some people use it as a top cover "heat" feeding material but are prepared to remove it at a moment's notice if the weather warms sufficiently to cause trouble.

"Sweet" feeds of various kinds are available and are excellent in special applications. One very coarse kind is prepared especially for horses and is not usually very effective as it consists largely of whole or cracked grain, rather than fine ground grains.

Sweet feeds are usually referred to as "percent" feeds because they consist of 10%, 18%, 20%, or 24% protein and are dairy feeds consisting of brans, ground grains, and alfalfa or timothy meal, with a covering of molasses. It also varies as to molasses content, some have ten percent or 18% molasses and so on. The molasses is a product of the sugar cane and beet industry and has a very high sugar content.

Sweet feeds should only be fed in outside beds but may be fed safely during warm weather and if the sugar (it converts to acid immediately, almost) is not allowed to build up it is a highly satisfactory feed in fattening or propagation beds. It is one of the better straight feeds (no tub mixing required) for adding to trenches in the beds for separating worms. It is used by preparing the trench then dumping in a row of dry feed, covering it lightly with bedding, then watering lightly daily until it is eaten by the worms. It also works very well as a mixture with corn chop for "core" feeding or heat feeding in the center of the bed, it should never be used as a heat feed on top because the sugar will be washed into the bedding and cause trouble there.

Sweet feed may be used as a "scatter" feed, that is, by scattering or sprinkling it on top of the bedding after the beds have been watered. The worms eat and convert the sugar to energy before it has time to sour. It should be scattered lightly over the surface and should not be fed more than once or twice a week, between feedings of other non-sugar feeds.

The list of feeds that may be used is obviously too long to be considered individually. We give here some of the more common or better types and explain mistakes to avoid with them.

The "mashes" are those feeds usually considered to be prepared only or especially for poultry. There is laying mash and growing mash. This might be for hens, turkeys, geese, ducks, and game birds that are grown commercially. There is little difference between the different kinds except in size of the ground particles. Growing mash is usually a little finer than laying mash. Game bird mash is usually a little coarser than laying mash- but this depends, to a considerable extent, on the individual manufacturer and his methods.

A Georgia feed manufacturer is furnishing a number of worm growers with a finely ground mash that appears on superficial examination to be various types of prepared cereals that have been picked up in stores because of over age or insect infestation and reground and rebagged for sale to worm growers.

We cannot quarrel with the manufacturer's product, no matter how it is acquired or what the ingredients because it is a good feed. It is however, far more expensive than other feeds of a like quality except for being coarser quality. We are now obtaining a "reground" lay mash from a Lakeland, Florida feed manufacturer. It is a high quality 20% protein feed that is run through the grinding mill the second time, and is about as fine in consistency as it is possible to grind without removing the bran or grain shell. And it is $.65 per hundred pounds cheaper than the *worm and cricket feed* sold for that particular purpose.

A reground worm feed in the from of lay mash or growing mash is available from any feed mill if it can be purchased in quantity by the grower. Naturally the feed mill operator cannot afford to run one or two hundred pounds--but it will pay him to run a ton, or even a half ton. We pay $.25 per hundred pounds for the regrinding and rebagging, although they use the same bag.

Lay mash or growing mash, then, is a "standard" feed in the worm industry, at least as near standard as may be in an industry that is to date completely disorganized and unregimented--if it ever changes I will start growing something else.

Lay mash, or mash, should be fed on the surface, as should practically all types of feed for all types of worms. Surface feeding encourages worms to remain near the surface for picking, and worms will do better near the surface where oxygen is plentiful. Surface feeding has all kinds of advantages for the grower, for one thing it saves him a good bit of time that may be spent in raking in feed in order to mix it with the top inch or so of bedding. It also may prevent damage to the worms from acid as the old method of "washing in" feed with a stream of water leached out many of the minerals and vitamins and carried them away, as it carried starch enzymes to the depths of the bedding where they could sour.

The advantages of surface feeding may not be apparent at first as worms must be trained to come to the surface for feed after they are accustomed to finding it everywhere mixed with the bedding. As the deep feed becomes exhausted they will rise closer to the surface as they know that is where the feed should originate. After growing accustomed to finding it there they will feed every night. This is true of all kinds of worms, even those that are not "surface feeders" normally.

The mashes are fed by spreading them on with the hands. It should be well distributed and should cover thinly, not blanket-like or in piles. A good example is a quart for ten square feet of bedding surface. This may look thin to some people who are accustomed to only feeding their worms once or twice a week. We advocate daily feeding at least during warm or hot weather. The worms should be watered BEFORE feeding so that the feed will absorb the moisture from the wet bedding below and this upward movement of moisture will also attract worms to the surface where the feed is.

Many types of grain, in fact all types in common use, are adaptable as surface feed for worms. They should be ground and preferably there should be a mixture of different kinds of grains. Very few grain products are a complete feed in themselves. Oats being one of the exceptions. Corn is fair but needs some additive such as alfalfa leaf meal in order to increase the protein content. Crushed or pulverised oats can be used alone to produce worms, but it is expensive and is better if other grains are added.

One of the neglected worm feeds is dehydrated alfalfa leaf meal. Every worm bed should have a good feeding of alfalfa leaf meal once a week during the summer. It should not be fed in winter or in closed buildings at any time except as a core in the center of the bed, and it may even cause trouble there.

Stick to those things easily available such as mashes and meals and if possible have them reground, for the finer the feed the less trouble it is likely to cause. The worms, if not overfed, can and will eat fine feed directly and digest it, but they cannot digest large particles of anything. It is the rotting of the large particles in most feeds that causes acid troubles.

It may seem that I am here refuting some previous opinions that I have expressed in other books regarding feeds, their value, and methods of use. I am refuting those opinions to some extent. Remember that I am a much older and more experienced worm grower than I was then, and many more things are available to the average worm grower than were available then. I still don't know everything by a long shot--but I know a lot more now than I did ten years ago, especially about worms.

I am not going to list here any of the complicated feed "formulas" that I have used. Literally any mixture of any vegetable or grain product is good worm feed. Some may be a little better than others but on the whole any of them are good if they are used properly. I stress proper use rather than the product you use.

As for the vitamin and mineral additives and mixes available to you, you can shop the grocer's shelves for many of them. The animal and poultry feed supply houses have many more, at cheaper prices. The most common good additive in the west is Tonax. In the south it is Dr. LeGear's Plus. In the northeast and east numerous feed supply houses maintain excellent stocks for all kinds of supplements. Anyone can write or go to the nearest Montgomery Ward and Co., or Sears, Roebuck and Co., store and request their current farm catalog. These list many types of supplements and antibiotics. Geritol works very well.

The best method of applying vitamin and mineral supplements to the bed is with a salt shaker. The amount will depend on what you are trying to accomplish but in no case should you use more than two ounces to each ten square feet of bedding surface.

Small operators can mix the supplements with the feed. Here about two tablespoonfuls to the gallon of feed will accomplish any result that larger amounts might do. A large application of supplement of any kind is wasted as vitamins are drained off by the action of air, sunshine and water. Minerals are leached out or washed away by water, thus it may be said that no vitamin or mineral is left after six or seven days, no matter how much was put in.

Some of the more exotic feeds are peanut meal, nut meals of different kinds available to people who live in areas where nuts are produced or processed, various other seed meals than any listed here. All will be good if used properly--and carefully.

Fruit and vegetable pulps should be used as a core type feed rather than as a surface feed although if dried and broken into small particles may be used on the surface too, where they may save many pounds and thereby many dollars, of ground grain feeds.

Generally the pulps of citrus fruits are a poor worm feed although they, too, are used in center cores, and to some extent the dried pulp is used for surface feeding. Such pulps should be fed carefully as they are coarse and require time to become food for the worms.

Sugar beet and cane pulp have never been a satisfactory feed for me. Cane pulp is a fair bedding additive if nothing else is available but it may heat badly or it may not heat for several weeks then suddenly heat for no reason, therefore it is not considered dependable. Sugar beet pulp takes too long to break down and also heats to some extent.

We will repeat here that any vegetable or grain product that can be cut, broken, or ground to a fine meal can be made into a satisfactory feed if properly used.

All feeds of every kind intended for surface feeding should be scattered by hand. Trying to throw it from a bucket or container results in piles or lumps that may sour or otherwise cause trouble. A little practice will enable anyone to scatter feed evenly and in the quantities desired. I and the man who feeds my worms can even spread to make a gallon, a bushel, or any given quantity cover a given area.

Never put down feed heavily over an entire bed. It should be sown thinly at all times with feedings given twice a day, if necessary, rather than to overfeed just once.

****

FEEDING IN BUILDINGS

The problems of feeding worms while housed in shelters or buildings should be approached with a pessimistic attitude on the part of inexperienced worm growers. Assume that something will cause you trouble sometime soon, especially if the building must be tightly closed during cold weather, and you will very likely be right.

The main trouble with inexperienced people is that they don't know what to look out for. The only sensible answer to this is: Everything.

The first thing that will probably happen is that the worms will either refuse to eat, after the first few days or weeks, or will refuse to eat and crawl to the top of the bedding and squirm about and die--or they will clean up all the feed in sight and still crawl to the top and squirm about and die. And they will die by the thousands. It is unlikely that one could lose ALL his worms in this manner but it does very often happen that one can lose so many that the season, and the grower, are ruined.

We use the same feeds inside that we do outside, with some exceptions: We never use sweet feeds, snapped corn, any of the coarse pulps or green leaf meals. We stick strictly to fine ground lay mash, and still have enough trouble to make us look for something better. So far we have not found it...except the peanut hull mix.

The bedding material used in inside beds dictates, in many instances, the type and amount of feed. We have found that the soilless beddings made from our formula are best for an inside winter operation. But Florida peat runs it a very close second. In

Florida we use a good grade of local peat moss obtained from the bed of a lake. This type of peat moss is available in other states, in most states, in fact, in the north and east, and we recommend that it be used wherever it is available cheaply enough (top price should be about $7.00 per yard delivered). The local product is always better than Canadian or German. In fact, nearly anything is better than the imported peats.

In the soilless beddings and the various kinds of peat moss you must use more feed than is normally used with other things. Cotton motes or cotton gin burrs and waste for instance, contain considerable food value in themselves. Soilless beddings and peats do not.

In order to correct some wrong impressions previously given in good faith let me say here that I recommend and use either soilless bedding or local peat of one kind or another for any inside propagation or fattening of worms. Any kind of worms grown commercially. The wild worms still do better in other materials covered by another book.

The use of peat then assumes reasonable heavy feeding while the worms are housed. Under most conditions the amount of feed is reduced drastically. The reduction should not be in the form of less frequent feeding but rather in reduction of the amount of feed given daily. The beds should be turned, or dug, at least once every two weeks, oftener is preferable, and every opportunity should be taken on warm days to open the building and allow a free flow of fresh air. If the air is not moving or no wind blowing, set up a fan or fans in order to force a good flow of fresh air. It is better if the beds can be turned while the air is blowing over them.

My own inside beds are watered daily because they are directly on the sandy soil earth, with no bottom of any kind and water drains off quickly. Beds with a cement or other solid bottom may not require daily watering during the time the building is closed and the flow of air is restricted, thus limiting evaporation of moisture from the bedding. The proper amount of moisture must be judged by the individual grower to meet his individual requirements. In buildings having several tiers of beds built on different elevations in the same room, some beds might require more frequent wetting than others, depending on their elevation and the amount of evaporation involved. If a fan is used to distribute warm air, for instance, the evaporation rate will be greatly increased by the movement of the air, thus more frequent wetting will be required.

Worms should be allowed to revert to a smaller size during winter storage. Feed should be reduced to almost subsistence level in most beds with a few beds, if necessary, used to grow out those worms needed for sale during such periods of storage. The beds reserved for growing out large worms for immediate sale should be those most favorably located as regards heat and admittance of fresh air. Collared worms should be picked or sorted from the storage beds and transferred to the fattening beds, rather than attempt to fatten out whole storage beds.

During exceptionally cold periods of time when we cannot open the buildings to admit fresh air we regularly treat with calcium carbonate every two weeks and once each month with Terramycin (Terramycin antigerm formula No. 77 by Pfizer). The Terramycin application is generally light, about one ounce for each 25 square feet of bedding surface. If we see any dead worms between scheduled treatments the beds are turned and treatment given anyway. If any appreciable number of worms appear squirming about on top of the bedding we treat heavily, wait three days, turn, wet, treat again with both calcium carbonate and Terramycin used in the second treatment.

The beds are fed only six days a week in winter. The seventh day, usually Sunday in my case, I leave them strictly alone with neither feed or water. This gives them a chance to clean up any accumulation of old feed that might sour or give trouble.

On exceptionally cold nights the air in the building might cool to the point that the worms will not feed properly. In this case do not water or feed. If water is put on, the old feed from the previous day's feeding will be washed into the bedding and is an invitation to trouble. Several days of cold may often aggravate this situation and the worms refuse to feed until the surface bedding is too dry for the worms to come up. In this case you must water anyway as lack of moisture will kill your worms as surely as any condition that may be induced by washing feed into the beds. It is a good rule in a case of this kind to apply a treatment of calcium carbonate immediately after wetting the beds.

Always, when wetting the beds during periods when the building must be closed tightly, we merely sprinkle the top of the bedding in order to dampen it thoroughly. We do not soak the beds heavily. In beds which have no artificial bottoms it is a good practice to water heavily occasionally so that some impurities may be washed away into the soil. Beds with artificial bottoms, should of course, never be soaked unduly as the water induces the formation of gases and acids and helps to trap them in the bedding.

One type heater not used enough by people in northern climates is the "blow-through" kind. This is a heater with a built in fan that pulls cool air in one side and blows out the heated air on the other. In Wyoming I used a natural gas heater of this type and this permitted induction of fresh air whenever it was needed. I merely placed the heater in an opening of suitable size right in the wall of the building. When fresh air was required I used the heater fan to blow it in and heated it during induction. A plastic cover was used to prevent entry of cold air when the heater was not in use. The heater itself being mounted flush with the outside wall.

Remember when raising worms inside on any scale (my buildings cover more than 7,000 square feet at present, with plans calling for another 10,000 feet) to water lightly, feed lightly, and turn and treat often, no matter if the beds do appear not to need treatment, treat them anyway. The treatment can do no harm and might prevent a lot of trouble.

Outside winter bedding and feeding of worms presents problems that are serious or troublesome in direct proportion to the severity of the climate. Usually the northern severe winters give less trouble than southern winters. In the north while severe winter cold covers the land, which is most of the calendar fall, winter, and spring period, the worms are completely dormant and very little care is needed. In the south winter comes and goes a dozen times during the winter season, trouble is almost a certainty unless beds are carefully and properly cared for.

Many growers use various products, especially the chopped straws, snapped corn, natural manures, cotton seed hulls, and waste vegetable products, for winter "heat" feeding in the south. Unpredictable cold and hot weather sometimes follow each other closely, and the temperature range may be from well below freezing to 70 degrees, or even higher, within a few hours. These sudden warm spells call for instant action when heat feeding using most of the commonly accepted methods.

Most growers use a cover of whatever it is they use for heat feeding. They prepare a deep bed, or allow the bed to accumulate more than the usual amount of bedding. Then they cover with an inch or two of manure, cottonseed hulls with a little cottonseed meal mixed in, or some green canning factory waste. This is then covered with several inches of straw. Sometimes a burlap or other cover is put on over the straw during extreme cold.

Some people follow radio and television weather programs carefully and act accordingly. Others depend on intuition and experience. With modern day weather forecasting, most forecasts are dependable and accurate, but the listener or viewer may live in an area remote from the point of origination, so far that his area may be more or less affected than is expected. This calls for some forecasting on the part of the grower. In some instances radio and television stations do not maintain up to the minute equipment and forecasts are late, therefore the weather may be upon you before it is announced.

In any case, if extremely warm or unseasonal weather is expected for more than one day, the heat feeder must remove some or all of the heating material or he stands a good chance to lose some of his worms. This is especially true of the common methods now in use.

By using a "core" of heating material down the center of the bed placed as described in another chapter and controlled with a top cover of hay or straw the grower saves a lot of work. During extremely warm periods the hay or straw can be removed with a pitchfork and piled for reuse when it turns cold again.

The core will probably need to be replaced about three times during the winter as its fermenting ingredients are exhausted. A cheap thermometer will help to keep track of the heat and heating qualities. In my experience corn chop with the peat moss mix is one of the best core mixes for winter heating anywhere, north or south. Good quality timothy hay is more effective to retain heat than straw, and burlap is a good cover. If desired you may also use tarred roofing felt alone or with the burlap. Burlap is manufactured in rolls, or bolts in different widths, and is not very expensive.

Some growers who use lumber, or boards, for making beds also use the same material for building up beds in preparation for winter. By adding boards they increase the depths of their beds to three or four times their working depth, or normal summer depth. Then they toss the contents of several beds into one, thereby making the same amount of heating material accomplish three times as much. The fact that the worms are very crowded does not seem to make much difference to the worms as the heating material supplies their food, and there must be plenty of it.

This works fairly well in African worm beds, too, but the favorite method is to use just plain cow manure in outside beds, covered with hay and a strip of burlap. The cow manure must be replaced fairly often, every week or two, so it is expensive and a lot of work as well. The main problem in Florida is to find a reliable source of cow manure, especially as it must be fresh in order to furnish heat.

The instructions and information given here and in other chapters work as well for all types of worms. Different conditions or changing conditions might change your problem. On the whole your problems will be no different than those encountered by other growers. Naturally explicit instructions cannot cover the broad spectrum of different types of worms in the different climates involved.

****

HELPFUL FEEDING INFORMATION

In this section I will attempt to cover some items previously mentioned or described but not detailed to the extent that some readers like. Most readers of any instruction book will choose some part of the instructions that appeals to their sense of values and will give much thought and time to working out the details of that particular section. I suggest that the foregoing chapters be studied carefully. There may be many things that you do not agree with and there may be some things that you already carry out in your own operation, but a careful study may give you new ideas for accomplishing whatever it is you want to do.

The average worm farmer is an extreme individualist. If he were not he would not be a successful worm farmer. I can put it another way: The average successful worm farmer is an extreme individualist. Being individualists, the growers are not so much concerned with trying to do things another man's way as they are in knowing how the other man does it. The successful worm farmer is not going to do things my way simply because I describe it in a book. But the grower who is having problems, and they all do, at some time or other, is going to look through the book for some ideas that might help him solve his problems. I want to help the man who wants to be helped, and entertain the man who wonders how the other fellow does it.

To help the man who wants to be helped, we have repetition. To entertain the man who wants to be entertained, we have different ways of accomplishing the same result...And it could be that even the individualist will try out one or two new things that he is sure won't don any good, but they are new.

Worms have habits, jsut like people, they develop habits that change to some extent with the climatic conditions and prevailing atmospheric conditions but they are well defined habits that can be put to work for the grower who is observant enough to map these habits and take advantage of them.

Most worms come to the surface to feed most of the time. They may not come out and crawl about on the surface, but they do come to within a few inches of the surface of the beds if the bedding is properly damp and is fed on the surface, as all beds should be. Some worms, notably the African nightcrawler, remain near the surface at all times, depending to some extent on the moisture conditions prevailing in the bedding, and in the air above the bedding. If the air is humid and moist the African worm will come out on the surface at night and crawl about in search of feed or a mate, or both as the case may be. This applies mostly to adult worms, or those large enough to be picked for bait or breeders. The young worms of most species remain deeper in the bedding than adults, and their size slows their movements to some extent.

"Hybrid" worms, manure worms, and all members of Eisenia or Helodrilus families (which account for practically all of the "commercial" worms) feed either on or near the surface of the bedding during warm weather. The younger worms remain deeper, as a rule.

Thus by changing their feeding habits we can take advantage of the worm's attempt to follow a pattern, whether that pattern is governed by instinct or some vestiges of intelligence is beside the point. Here is a chance to prove that we are smarter than the worms.

Our own method of separating worms in the beds takes the following course: We rake up the bedding from the sides to the center, just rounding it off gently, not piling it steeply. Then we cut a trench in the top of the heap by using a rake to pull a small amount of bedding back, leaving the trench about three inches deep. Then we water this with a sprinkler, getting it soaking wet, so that water can be squeezed from the bedding when a handful of it is picked up.

Using a wash tub (or a contractor's wheelbarrow or other suitable portable container) fill it 2/3 full of corn chop (corn, cob, husks, ground into a coarse feed). To this we add two gallons of dried whey (a dried milk product that can be obtained from any feed store) which is usually dried skim milk, though some types still have 1% to 2% of crude fat (butter) remaining.

We add enough water to this to make a thick paste which is mixed thoroughly to distribute the dried whey into the corn chop. Do not add enough water to completely dissolve the whey, but only enough to make it pasty, or gummy.

This paste is then used to fill the trench previously prepared in the bedding. Then we rake some of the bedding back over the row of paste in order to keep out flies and other insect pests as well as to keep the paste from drying and caking. We do not water at this point, but we do feed some lay mash, by spreading it thinly on the sides of the mound of bedding. We try to prevent getting too much feed on top of the row of paste as the large worms will follow it and tend to uncover the paste by rolling the bedding down the sides.

Give this three days, watering lightly and feeding daily but feeding only the sides of the mound.

On the mounded beds the small worms will gravitate to the center and beneath the row of feed previously distributed (the paste) in the center trench, while the larger worms will gravitate to the sides in order to avoid both the small amount of heat dissipated into the bed by the fermenting paste, and the concentration of small worms. Large worms, especially fat large worms, do not like to compete with young worms for feed, and will usually move away from heavy concentrations of feed where their size permits them to move faster than small thin worms. This free movement allows them to obtain plenty of feed to maintain their size and put on more fat.

When picking these mounded beds the pickers should avoid digging into the row of feed. Most of the large worms will be found just under the surface of the bedding and the bed can be reshaped by merely smoothing and re-rounding the sides of the mound.

Mounding, incidentally, is a practice that should be used by growers of all types of worms. The rounded bedding offers more surface area for the absorption of water and especially air. The latter is most important. Large fat worms require more oxygen than thin young worms and a free movement of air is essential to growth and well-being of all worms.

We round up beds even in the buildings during the winter. We do this by first digging down the center and loosening the bedding there with a pitchfork. Then with the same fork we scoop up the bedding on the sides and turn it upside down on top of the loosened center. We use a six tined fork for this and do not rake or smooth the bedding afterward but leave the lumps and bumps. The worms and the action of water as the beds are moistened will soon have them level again, at which time they start to pack and need to be turned again.

BETTER PROPAGATION

As explained in another chapter, feeding for propagation or fattening is a double-barreled problem. When we feed to fatten we don't want propagation but when we feed for propagation we do want some fattening if possible so that the larger worms can be picked and sold to pay for some of the expensive feeds being poured on the beds. Usually any feed that is suitable or efficient for fattening worms is also the best type to increase propagation. For this we have found no answer, except by chemical means.

For both increasing size and propagation activity we have found alfalfa meal to be almost at the top of the list. In beds of hybrids and manure worms, all of the Eisenia and Helodrilus families, manure of one kind or another, with cow manure topping the list, increases propagation more than any other type of feed tested, with some exceptions. In hot summer weather, either fresh or aged manure might decrease propagation by increasing the temperature of the bed to such an extent that the worms estivate. On the other hand, during spring, fall, and winter, the heat engendered by liberal applications of fresh manure increases breeding activity to an astonishing degree. African nightcrawlers will breed faster in pure fresh manure than in any other type of bedding or feed, and this holds true for either summer or winter.

However, there are many times and circumstances where manure, for one reason or another, cannot be used. In some areas where Africans are raised it is almost impossible to obtain manure, either fresh or aged. In other areas it is very expensive. Some growers have their operation so located as to offend neighbors, or come athwart the law, if they use any kind of manure. Manure of any kind, except the light applications of synthetic manure described in this book, does attract flies which produce maggots in the beds which hatch into more flies, thus hatching trouble with any neighbors and sometimes the pickers.

My own solution to this problem was not easy, but the application was easy once I learned. I use alfalfa leaf meal. This is the ordinary kind obtainable at most feed stores and supply houses. Some leaf meals are 17% protein, some more or less, and there are other variations such as those with molasses added, sugar concentrate added, etc. The kind I use for inducing propagation is just plain 17% protein alfalfa leaf meal.

Alfalfa leaf meal is fed like any other feed. My own method of feeding is worth repeating here because it is the result of many years of research, application, and study of the results: I water the beds first, keeping them good and wet at all times. It is my opinion that most beds remain too dry for good propagation and for this purpose it should be wet enough that water can be squeezed out of it. Of course this should be approached with some caution by people who have cement or other waterproof bottom in the beds. It is possible to drown young worms and render infertile some eggs by too heavy watering of waterproof beds. In waterproof beds the bedding should be turned to check moisture content of the lower layers of bedding, and also to incidentally release gases and acids trapped there by the oxidation of feed and bedding.

Immediately after watering the beds they are fed by scattering a thin layer of feed over the entire surface of the bedding. During the growing season I increase propagation by alternating feedings of reground lay mash and alfalfa meal. One one day and the other the next, for six days. Sometimes this is altered to every third day for twelve days, with the grain mash every day in between, of course. Usually one hundred pounds of alfalfa meal will feed about the same area (square feet) as one hundred and fifty pounds of lay mash. This is because the alfalfa is finer, almost as fine as flour, and goes further as it appears to be more heavily concentrated on the surface. With our specially reground feed we get about the same surface coverage with one as with the other.

When the young worms are produced in the beds we change to a molasses or sorghum feed. This is a dairy feed composed of ground alfalfa (not as fine as alfalfa leaf meal), brans, and cracked grains. Various concentrations of this type of feed are available also. Some contain 13% molasses, some more. We use 18%. The amount of molasses is determined by weight. The syrup is sprayed on the feed as it is mixed and they add 18% of the weight of the grain particles in order to arrive at the percentage of molasses used. This type of feed brings on the young worms faster than anything we have found. A better method is to continue to feed the regular lay mash and alfalfa meal and buy the sorghum in bulk (five gallon cans) if it is available. We understand that molasses or sorghum is not obtainable in some areas without a great deal of trouble, while sweet feed, or molasses sprayed dairy feed is universally obtainable.

The bulk molasses are dissolved in water, one gallon to three of water, then is sprayed or sprinkled on the beds. There are "force" type sprayers that work in conjunction with a garden hose. These force type sprayers are cheap (they are used to dispense controlled quantities of insecticides, liquid fertilizers, etc.), easy to use, and adjustable so that quantity can be controlled.

The beds should be watered lightly for a few days after the molasses are used. This prevents washing this expensive feed through the bedding and into the ground before the worms can make use of it. In the case of force sprays, set it to get the molasses on with the smallest possible amount of water. A gallon of molasses (it varies some in sugar content and viscosity) should cover about 200 square feet of bedding and should be used every day for nine days (three feedings of it) then should be discontinued to observe the result.

A FOOTNOTE TO THIS IS THAT MOLASSES OR SUGARED TYPE FEEDS OF ANY KIND SHOULD NEVER BE USED IN A CLOSED BUILDING. THE FERMENTATION CAUSED BY LACK OF FRESH AIR AND SUNSHINE WILL RESULT IN SERIOUS DAMAGE TO THE BEDS AND THE WORMS.

Sorghum cane products are more expensive and difficult to obtain in the north west than they generally are in the south. Many areas of the south raise sorghum as fodder, and many farmers boil off their own molasses. Either the commercial kinds or the home made kind is fine for all kinds of worms although home made kinds vary more in sugar content.

I may say here that one of the finest methods of starting a native (northern) nightcrawler bed, or of increasing production in an area where they occur naturally is by using sorghum molasses or any of the crude molasses available as a by-product of sugar mills, either cane or beet. Use one gallon to each one hundred square feet of area and dispense it as described above. It is particularly effective in spring when the worms first begin to appear on the surface, which is also the most active reproduction period.

One method of use for this purpose is to soak corn or wheat bran in the diluted molasses, then scatter the soaked bran. The worms are very hungry when first appearing in the spring, after a winter of inactivity, adn they will eat the bran in large quantities is it is made available for them. This increases crawling activity, which in turn increases their chances of meeting a suitable mate, which results in increased production.

Twenty pounds of bran will soak up half a gallon of molasses diluted with half a gallon of water. This method can also be used in beds for other kinds of worms and is fairly satisfactory, though a lot more work than merely purchasing "sweet" feed in the first place.

SOILLESS CAPSULE BED

Did you ever get an inquiry or order for capsules that you considered filling but refused because you thought it might be too much trouble? I have, several times. Then I got an order for ten thousand capsules at a far better price than I had ever gotten for worms, one that I could not afford to turn down. An opportunity to earn a large sum of money is one of the best ways I know to inspire some real thinking.

The result of my own brain storm has been the secret of making many, many successful shipments of capsules--and sales that I otherwise would have had to miss.

My method makes use of ordinary burlap. You can buy it in strips a yard wide and several hundred feet long. Or you can use several old feed bags that you might have on hand or be able to obtain easier and cheaper than you can buy the regular burlap bolt goods.

Fold the burlap to fit a square box about eight or ten inches deep. The box can be any dimension to fit the burlap that you happen to have on hand, or vice versa. Or to fit the need for capsules for shipping, for painting new beds, or for whatever purpose you might need them.

Make several folds, at least 6 or 8 or as many as 14. The burlap can be placed on the bare bin or box bottom, or it can have a little bedding under it. I used it without bedding.

Be sure the burlap is soaking wet throughout. As the wet burlap is folded sprinkle just a small amount of lay mash on the flat lower side. Not too much of the feed, just a very light sprinkling--you don't want this to heat too much, a little heat may help, and be sure that the small amount is distributed evenly over the entire surface. The feed can be lay mash mixed with crushed limestone, one part of limestone to ten of feed, or chicken manure, well pulverised, into which one part lay mash has been mixed, together with half cup per gallon of crushed limestone. Or pulverised rabbit manure will work very well here. Crushed oyster shell will work as well as crushed limestone.

Each layer of burlap should be sprinkled lightly with water after the feed is distributed on it. be certain that the burlap as well as the feed is well soaked but not drippy, running wet.

When all the folds are made, one above the other, let it soak overnight. Place the hand between some of the bottom folds to be sure that it is not heating too much, it will heat a little. If it is heating insert some sticks of wood between every third fold so that some air will get to it. It should stop heating within three days.

If the capsule bed is not heating, or after it has cooled, fold the whole thing back in one heap. It may be quite heavy, but you must start at the bottom for the next step. Introduce some of your largest and oldest worms into the folds, starting with the bottom one and replacing each fold as the worms are placed in. Place about 100 worms to the fold if it is about 4X4, fewer if it is smaller, more if larger. Be sure the burlap and feed are soaking wet completely through all the folds before the worms are introduced.

The worms will work back and forth between the rough weaves and from fold to fold to some extent, but keep it well watered and in a reasonably cool, but not cold place.

Leave the worms in the burlap for about five to seven days. Examine the feed in the center folds occasionally, about every three or four days. If the feed is all eaten, add a little more by laying back all the folds then working the burlap back in the same manner as when first starting. Be sure none of the worms crawl out while you are doing this. It is possible that during the short time the bed has been in operation that there will be some capsules. Sometimes it occurs in three days or four, sometimes it takes longer. Weather, humidity, and many things enter into it.

After ten days the worms will be much larger, unless you started with exceptionally large worms to begin with, and they will be cleaner and tougher. There will be hundreds, even thousands, of capsules, and perhaps some young worms in the folds. These capsules should not be touched with the dry fingers as body moisture and oil sometimes causes them to mold. They should be carefully and gently scraped together with a plastic ruler, a piece of cardboard, or other non-metallic tool. If you can refrigerate the capsules at about 36 or 38 degrees they will keep for several weeks, if you want to gather and accumulate them for that long. If you ship capsules this is the easiest and cleanest method of getting them out.

Capsules can be air dried and preserved for long periods of time but this is too tricky for the average grower.

If there are lots of young worms left in the burlap, spread it out over a bed and allow it to dry there, and the young worms will crawl into the bedding to avoid dehydration, after which the burlap may again be used for the same purpose. it will last for five or six uses, at least.

The large worms collected serve two purposes. They will sell at a premium rate because they are conditioned and toughened. If you will advertise capsules there is always a market for them, and at premium prices. Most growers ask and get eight to ten dollars per thousand for capsules.

There are several successful methods of shipping capsules. During the warm summer weather it is almost impossible to prevent them from hatching in shipment (unless you learn to dry them and pack them in dry cotton), at least some of them. These tiny worms can crawl through very small openings so it is advisable to ship capsules in a paper bag, folded over TWICE at the top and stapled several times so there is no possibility of the newly hatched worms escaping. The best bag to use is the double wall R.C. Bait Bag.

Use a small ball of damp cotton about as large as your two fists. Open it up, so that it is cupped, as a bird's nest, Place the capsules in this nest then fold the tops over so that they will be in the center. Place this ball and capsules in the R.C. Bait Bag, staple, then place the R.C. Bait Bag in a strong carton to prevent possible crushing of the capsules. Some people use a sealed glass container but the dampness may cause the capsules to mold, if there is not a free flow of fresh air. Of course a pint fishbait cup may be used if a layer of dry cotton is used at the top. The dry cotton will prevent the hatched worms from crawling out through the ventilation holes.

The buyer should know what to do with capsules when they are received. The ball of cotton, with the capsules inside, should be buried in warm bedding that has been prepared well in advance to prevent any possibility of heating. A small amount of bedding should be raked over them to cover enough so that they will remain damp during the incubation period. They will hatch in from three to seven days if the temperature is above 65 degrees.

Many people have asked me what they should charge for capsules. Capsules are valuable to the extent of the work that has gone into gathering them. By using this method you can sell them cheaper than usual and earn a very good profit. I formerly sold them at $9.50 per thousand, postpaid. I only produce African worms now and do not sell capsules as they sometimes hatch within two hours of production, making it impossible to handle them satisfactorily.

Inasmuch as the customer stands to obtain in the neighborhood of six thousand worms from the hatched capsules, depending somewhat on the age of the breeders that produced the eggs, it is a very good buy for both buyer and seller. Capsules from younger worms will produce fewer young than capsules from old worms. Very young worms, just beginning to produce capsules, average fewer fertile eggs per capsule, sometimes only two or three. On the other hand, two or three year old breeders will produce capsules with from seven to thirty eggs--and more of them will hatch and live. It is safer, if you do not know the age of your breeder worms, to advertise four or five worms to the capsule, and be on the safe side. If there are more it hasn't cost you anything extra and the customer is that much happier.

More of us should advertise and sell capsules, especially of "hybrid" varieties. It is a very satisfactory method of either buying or selling worms.

FEEDING IN THE PACKAGE

Many people have written to find out if I have worked with or solved the problem of feeding worms in the package after they are prepared for shipment or storage on dealer's shelves.

I have never considered this a pressing problem. Worms packaged in ordinary peat moss (I use and recommend Canadian peat in either one of several brands that are available at feed houses and gardener's supply stores) will keep for long enough to ship in good condition from coast to coast or from Florida to most parts of Canada or Mexico.

Worms packaged in retail size packages that are several days in shipment, then may expect to remain on the dealer's shelves for another several days may, in fact should, be fed in some manner before leaving the worm farm.

There are several satisfactory methods of doing this and all depend, more or less, on the climatic conditions prevailing at the farm or the point of arrival, or both. Geography plays a tremendous part in the distribution of worms throughout the world, and plays a very important part in the condition of commercially produced worms both at point of origin and point of use. This is of course especially true of worms confined in large numbers to a small space.

The most common, and easiest of methods is by the use of oil-soaked peat. Not just any kind of oil and not really soaked. About two pints of oil to a standard 6 cubic foot bale of peat will feed a large number of worms for several days and will at the same time inhibit evaporation both from the peat and the skin of the worm, thus serving the purpose of keeping the peat moist longer and keeping the worms larger by preventing loss of moisture through the skin.

Peanut oil is the only completely satisfactory type of many tested. Corn oil is fair under some conditions of cool and damp weather, but may slow evaporation of moisture and prevent cooling in hot dry weather. Corn oil makes a tighter harder film than peanut oil.

The best method of using peanut oil (it is available at any well stocked grocery store) is by preparing the peat as usual--presoaking and "fluffing" by rubbing it between the hands. The fluffed peat is then placed in a tub or other suitable container and is packed by pressing it down with the hands.

Over the peat thus packed in the tub (the tub will only be about half full of packed peat) pour one half pint of peanut oil. Pour the oil around on top, not all in one place, and allow this to soak overnight. The peat should then be further prepared by refluffing it and adding any water that may seem necessary. This is satisfactory for packaging bulk worms for long journeys. To prepare pints or small packages for resale, use three quarters of a pint of oil in the tub of peat. The measures are not critical and a little more or less will do no harm. Except in real hot dry weather be sure that these measures are not exceeded. Too much oil may inhibit evaporation to the point that the worms would be better off without it.

Another method is the use of corn protein. This is a product used by manufacturers of binders (glue) and of ink, where it is used as a hardening agent to make the ink stay on the paper. Two tablespoonfuls of powdered corn protein mixed thoroughly with one quart of pulverised sewer sludge will feed many worms for a long time. This is best dispensed with a large salt shaker into the packages after they are prepared, the feed on top, never mixed. About a teaspoonful for each hundred worms (Eisenia red worm).

If sewer sludge is not available the corn protein may be used with calcium carbonate, but is not so satisfactory or efficient. Of course the protein may be mixed with any other type of material that will not heat, even a small amount of rich top soil or dried and pulverised field manure. A teaspoonful of pulverised field manure and corn protein will feed a hundred worms for a surprisingly long time. Corn protein may not be generally available everywhere but is manufactured under the trade name Zein, by Corn Products Manufacturing Co., Pekin, Ill.

The most efficient method yet found for feeding worms in the package is by the use of Buss Bed-ding, a product manufactured by Buss Manufacturing Co., Lanark, Illinois. It is a fair worm keeper when used as directed but is most useful, in my opinion, as a packaged worm feed.

Buss Bed-ding should be mixed as directed on the package. As the worms are packaged, using regular peat moss, or whatever you use, place a layer of Buss Bed-ding on top of the peat. The Buss Bed-ding should only be half an inch or so in thickness and may be permitted to cover completely, like a mat, or it may be just scattered or sprinkled in the package on top of the peat. It should be wet, of course.

Buss Bed-ding may also be used by mixing it with the peat when the peat is prepared for use. In this case use only one part of Buss Bed-ding to nine of peat moss. A few of the people who tested this product claimed that it shortened the life of the package. Most packages will last longer than the worms so a slight reduction in package life would not be very serious.

THE END

ISBN 0-914116-02-9